Praise for
On the Run

"[A]n original debut. Informative. . . . Entertaining. . . . Fun and refreshing." —*Publishers Weekly* (starred review)

"A fantastic book. . . . A delightful tale vividly brought to life by DiBenedetto's lucid prose."

—James Prosek, author of
Joe and Me and *Fly-Fishing the 41st*

"Combines the best elements of fishing journal and travelogue."
—*Los Angeles Times*

"This book deserves to stand alongside efforts such as John N. Cole's *Striper*. . . . Sprightly, sensible, and engaging."
—*Library Journal*

"*On the Run* ranks among the finest outdoor writing I've read in the last decade. It's hilarious, articulate, lyrical, and real. The first chapter alone is worth the price of passage. But you won't stop there."

—Randy Wayne White, author of *Everglades* and
The Sharks of Lake Nicaragua

Amy Berkley

About the Author

DAVID DIBENEDETTO, a native of Savannah, Georgia, is the deputy editor at *Field & Stream*. He has written for *Men's Journal, Rolling Stone*, and *Salt Water Sportsman*. In his free time he can be found fishing for stripers on Long Island Sound and in the surf at Montauk, New York.

David DiBenedetto

ON THE RUN

AN ANGLER'S JOURNEY DOWN THE STRIPER COAST

Perennial Currents
An Imprint of HarperCollins*Publishers*

FIRST PERENNIAL CURRENTS EDITION PUBLISHED 2004.

Designed by Adrian Leichter

Illustrations by Keith Witmer

The Library of Congress has catalogued the hardcover edition
as follows:
DiBenedetto, David.
    On the run : an angler's journey down
the striper coast / David DiBenedetto.
        p. cm.
    ISBN 0-06-008745-5
    1. Striped bass fishing—Atlantic Coast (U.S.).
    2. Striped bass—Migration—Atlantic Coast (U.S.).
    I. Title.
SH691.S7D53 2003
799.1'7732'0974—dc21   2003050130

ISBN 0-06-008746-3 (pbk.)

04 05 06 07 08 ❖/RRD 10 9 8 7 6 5 4 3 2 1

FOR MY BROTHER STEPHEN,
WHO HAS BEEN THERE FROM THE START

Best of all he loved the fall.

—ERNEST HEMINGWAY

Contents

ON THE RUN

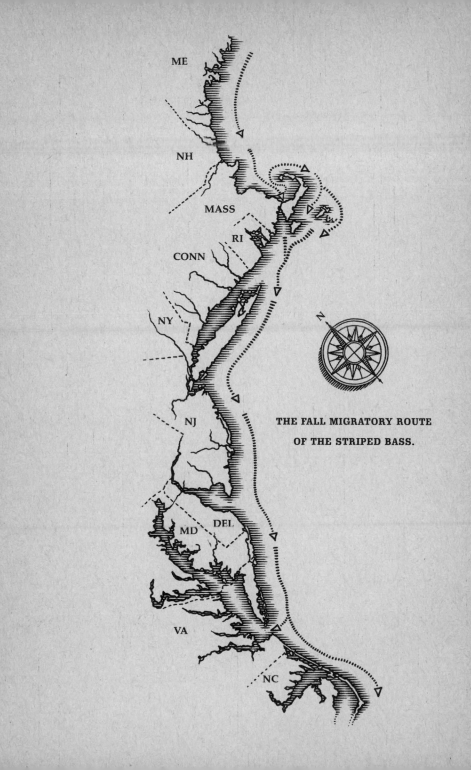

ME

NH

MASS

RI

CONN

NY

NJ

MD   DEL

VA

NC

N

THE FALL MIGRATORY ROUTE

OF THE STRIPED BASS.

Swimming with the Fishes

I don't advise peeing in your wet suit," shouted Paul Melnyk. "You'll get a mean rash."

"I'll keep that in mind," I hollered into the wind, my words quickly blown back into my face.

Truth was, whizzing in my neoprene bodysuit was the least of my worries. I was standing at the edge of the roiling Atlantic in Montauk, New York. Clouds covered the sliver of a moon, the chilly October night as black as the bottom of a well. In a few minutes I would follow Melnyk into the ocean.

We planned to lie on our backs, fishing rods held under our arms, and kick our way three hundred yards offshore. Once there, we would float on the current that ran parallel to the beach, casting live eels for striped bass. After we were carried for a half mile or so, we would kick back to where we started and begin the drift again. Melnyk, a Montauk local who invented this form of angling, calls it skishing (a cross between skiing and fishing, since large stripers often towed him like a water-skier). As one guide told me before my skishing adventure, Melnyk was on "the extreme end of extreme."

The night before our trip, a surf fisherman, with both feet on shore, had landed a fifty-pound striper, and Melnyk wanted to best that mark. He was sure the fish was an indication of a school of trophy striped bass in the area. There was no turning back. We huddled behind a large dune to zip our wet suits and run through an equipment check. Headlamp? Check. Whistle? Check. Pliers? Check. Knife? Check.

"If a shark grabs me, I expect you to fight him off with your knife," said Melnyk, trying to loosen up the situation. His levity was lost on me. I knew enough about the area to realize that the threat of sharks was no joke. A little more than a decade back, a Montauk charter boat had landed a monstrous great white (seventeen feet, 3,427 pounds) that had been snacking on a dead whale not far from the point, and just that summer a fourteen-foot mako had been pilfering stripers from the ends of fishermen's lines and ramming boats near Cape Cod. Up and down the East Coast, 2001 had been the summer of the shark.

There were also rip currents, some of which ran at eight knots. If we got caught in the wrong place, we would be shot out into the ocean as if on a water-park ride.

"Let's do this before I chicken out," I said.

"Okay. Remember, this is a shore break. It's dangerous surf. These waves can pick you up and slam you on the beach. It'll ruin your week. Once we clear them it's easy sailing—make that kicking."

We waded in. The surf zone was a cauldron of white water, and beyond it the sea's lumps melded with the sky. When the water reached our knees, we dropped on our backs and pulled on our flippers. With the waves rushing to shore, it was a clumsy endeavor, and twice I rolled face first into the water before succeeding.

"Ready?" yelled Melnyk.

"Ready."

"Kick, kick, kick, kick."

When my flippers gained purchase, I zipped ahead into the wash.

"Stay with me, Dave."

We were kicking side by side when the first breaker rushed over us. I swallowed a mouthful of seawater and bobbed to the surface. Melnyk hooted with delight. The next wave lifted me up and carried me tumbling back where I'd started.

"Come on, Dave. Kick, kick, kick!" coaxed Melnyk. I gathered myself and pushed off again. In less than thirty seconds we were out of the surf, rising and falling on the choppy waves of the ocean. At fifty-six degrees, the water breaching our wet suits was breath-stealing.

"Is this living or what?" screamed Melnyk. "We're on the edge, man."

"How much farther?"

"About a hundred kicks."

On the crest of each wave I could see the shore. TV sets were flashing muted blues and reds in the windows of the beach motels above the dunes. I found myself envious of the occupants, who had little fear of disappearing into the Atlantic. Suddenly there was a surface commotion just in front of our heads, like a broom slapping the water. "Just a flock of sea ducks riding out the weather," said Melnyk. "We probably scared the shit out of them."

"Likewise."

Eventually Melnyk yelled, "We're here, man." We stopped kicking. The thick wet suits provided enough buoyancy to keep us chest high in the gin-clear water. I flipped my headlamp on; it illuminated a circle in front of me, the strobe reaching toward the bottom. I could see my purple flippers flexing as I flutter-kicked to stabilize myself. I wondered what could see me from below, and inched closer to Melnyk.

Melnyk kept each eel in an individual sandwich-size Ziploc bag for easy handling. He passed one to me. It squirmed within its plastic confines as I hooked it through the jaw. With the hook in place I pulled hard, ripping the bag and freeing the eel. It danced on the end of my line. If I didn't cast soon, it would tie itself and the line into a slimy knot.

I turned my light off. Melnyk already had his eel in the water. He was floating about ten feet from me. His black neoprene hood combined with his surf rod rising from the surface made him look like a seagoing knight, his trusty steed a sea creature from Proteus's flock. As I went to cast, Melnyk's rod quivered, then bent deeply. He reared back on the fish, then yelled, "Oh baby, they're here."

III   III   III

In 1954 the editors of *Salt Water Sportsman* declared that the striped bass had "a fanatical human following that compares to none." If Paul Melnyk is any indication, the striped-bass fervor is stronger half a century later. It's estimated that on the East Coast more than three million fishermen pursue the fish. Feeding the passion is a striper population that has soared to a 150-year high after experiencing a near total collapse in the early 1980s. The reasons for the decline were legion—from overharvesting to pollution of the spawning grounds—but thankfully legislation and the resiliency of the striper helped reverse the damage. The stock resurgence was in full swing when I joined the striper faithful in 1995 not long after I moved to New York City to begin my career as a journalist.

Having grown up on the banks of a coastal river in Savannah, Georgia, fishing was as much a part of my character as my southern accent. I spent the days of my youth immersed in the pursuit of fish. The walls of my room teemed with photos of sea life clipped from sporting magazines: Sailfish herded bait with their electric-blue dorsal fins; giant bluefin tuna rose from the depths; redfish tails wagged above the surface. On the ceiling above my bed, next to a shark jaw suspended on a piece of fishing line, I tacked a chart of the local waters. I liked to study it before I went to sleep. One night my father opened my bedroom door. He inquired about my homework, then took a good look around. "You know, Dave," he said, "at this age your brothers had posters of women on the walls."

Instead of fast-food joints or lifeguard stands, I found work on the decks of private and charter fishing boats. Every time I sank the gaff in an especially green or wildly thrashing fish aboard the *Captain D*, a thirty-foot charter boat, my boss would turn to our party and say, with pride,

"That, folks, is a sign of a misspent youth." When it came time to choose between two colleges I wanted to attend in New England, I opted for the school surrounded by the best trout water.

In New York, I took fish where I could find them. On weekends, I fly-fished in one of Central Park's weed-choked ponds for stunted largemouth bass, and once foul-hooked a baby stroller on my backcast. Fortunately, the stroller's occupant was on a nearby blanket. While navigating around the city, I frequently took detours through Chinatown to look at the blackfish that swam in crowded tanks before being sold to passersby for chowder or whole roasting. In midtown, I stared through the windows of high-class seafood restaurants at the dead fish—snapper, grouper, monkfish, salmon, rainbow trout—laid out on beds of ice. There were also countless trips to the Coney Island Aquarium. These routines did not placate my needs, only exacerbated them.

Then, on a foggy night in Long Island Sound, I landed my first striper. It weighed twenty pounds and was just over thirty-six inches. That a fish of such size could be swimming in the shadows of New York City's skyline in a waterway I thought corrupt with pollution amazed me. I was, to borrow a term from the writer Charles Gaines, "gut-hooked."

Striped bass had not been part of my angling education, but I soon learned that the fish migrated from warmer climes to the waters surrounding the city in the spring and left in late fall. Their catholic appetite made them suckers for an array of lures and baits. Best of all, I discovered, I lived literally a New York minute from prime fishing grounds: the Hudson River. Often I would wake when many of my friends were returning from a night on the town, walk three long blocks from my apartment—across Broadway, West End Avenue, and Riverside Drive—to the Hudson, and land a handful of fish before work. I was usually alone, save for the brown blur of rats returning to their riverside lairs and bums sleeping off hangovers on park benches. It was during these sunrise expeditions that I began formulating the ultimate angling adventure. For years I kept the plot to myself, carrying it around like a lucky rock. When the absurdities of life reared up, I retreated to my plan, honing details, knowing one day I would need to execute it.

I wanted to follow the fall migration of this grand fish, for fall is the best time to be a striper fisherman. With the task of their southward movement burning up energy and the need to store fat for the approaching winter, stripers seem to do nothing but eat. It is a movable feast. And they're never far from shore. During my first fall in Montauk, I had watched this river of feeding fish pass by after a few weeks of inconceivable action. Perched on a rock in the surf, I had looked out on an ocean that was awash with life—stripers, bait, birds—and watched it go blank. Not a temporary blip but a blackout. On the stripers' tails, a northwest wind, cold and stiff, blew like stink for three days. There would be stragglers, but the giant game convoy had vanished, or so it seemed. Really, it has just moved farther south, to the delight of other anglers awaiting its arrival.

My journey would begin in Bath, Maine, near the northern end of the stripers' range. There I hoped to find the fish on the cusp of their fall migration. When they headed south, so would I. My path would intersect with the grand locales of striper fishing: the Kennebec River, Nauset Beach, Martha's Vineyard, Cuttyhunk, Jamestown, Montauk, Cape Charles, Oregon Inlet. I'd finally tap the brakes for good, some three months after I set out, on the Outer Banks of North Carolina. Here the striped bass spend the winter.

I would pursue the fish in the surf, from all types of boats, and even underwater with my camera. In doing so, I'd spend time in the company of the country's finest sharpies, as the best striper fishermen are called. If my timing was decent, I would keep pace with the convoy of fish.

With the trinket shops of summer-vacation towns shuttered and traffic jams replaced by leaves loping along grassy shoulders, the road would be mine. At night, when I wasn't fishing, I'd camp, bunk with friends, or crash in cheap seaside motels.

For some time the realities of my scheme, though assuaging, seemed insurmountable. But my obsession with the striped bass eventually wore them down. Seven years after landing that first striper in Long Island Sound, I chucked my job, tried desperately to explain the situation to my girl-friend, found a vehicle, and cemented an itinerary. I told my bewildered friends I had become a striper groupie.

III   III   III

With the fish hooked, Melnyk leaned back for leverage, as if sitting in a submerged BarcaLounger, his flippers poking above the surface, and went to work. He whipped the striper in less than five minutes and then grabbed the leader, steer-ing the twenty-pound, three-foot-long fish under his arm. A big striper, but not the cow he was hunting. "This one's as docile as a baby," he said, reaching for the hook. Unfortu-nately, the sea wasn't as calm. A large wave, its breaking crest alerting us it was on its way, lifted us skyward. For a second atop the six-footer, a gust of wind smacked us like a balled-up wet towel. "Whoa, who turned on the fan?" said Melnyk. Then we dropped down the back side of the wave. The heavy weather had me spooked, and it was only going to get rougher. I made a deal with myself: no copping out for at least thirty minutes.

A few casts later I felt a solid thump on my line and set the hook. Immediately, the fish began to have its way with me; I plunged face first into the water. I pulled my head out to hear Melnyk laughing but then went down again, my legs kicking like mad. This time I managed to lean back and exert some pressure. The fish charged for the shore, spinning me around and, again, yanking me off of my axis, but I recovered much quicker. Back in position, I began to get towed away from Melnyk. "Hey, don't let me get too far away," I yelled.

"Ha! You're doing great. Might be a thirty-pounder."

After ten minutes I had the fish in close. Then it came thrashing out of the water, its teeth clacking, head swinging back and forth. I was eye to eye with a ten-pound, pissed-off bluefish, one of the ocean's gamest, and dirtiest, fighters. There would be nothing docile about this landing. As the fish circled, I tried to reach out and grab it, but every time I did the blue rose up and tail-walked across the surface, snapping its jaws like a set of hedge clippers. "Watch those chompers!" yelled Melnyk. "Lose a finger out here and we'll have one hell of a chum slick." Finally I got a grip just behind the blue's head and pried the hook free. The catch did little to buoy my courage.

For the next twenty-five minutes, neither of us even registered a hit. "I think we need to go out a little further. Maybe another hundred yards," said Melnyk. "You up for it?"

"I think I might want to head in," I said. "I'm a little freaked out tonight."

"We can go in if you want. Just say the—Wow, what was

that?" His sentence had been interrupted by a deep grunting sound coming from directly beneath us.

"I was hoping you would have a good explanation."

"Maybe it's a blackfish marking its territory."

"Blackfish don't grunt."

"Just another one of the wonders of being out here," he said.

My wonderment ended a few seconds later when I told Melnyk I'd had enough. "You sure, man?" he said.

"Positive."

After taking a drumming in the surf on the swim to shore, I stood up in a few feet of water. My legs felt like wet noodles. When Melnyk dropped me off, I took a long, hot shower, then went to the Dock, a rowdy bar favored by fishermen, for a double bourbon on the rocks. I'd never been so happy to be on solid ground, oh-so-wonderful solid ground.

The Run

When describing why a proven drop may be hot one day and cold the next, seasoned anglers like to say, "Fish have tails." These fishermen understand that their quarry moves according to patterns of wind, tide, time of day, and myriad other variables. Fishing a good spot under the wrong conditions is a lot like lacing up your ice skates in July.

A striper's tail, or caudal fin, is shaped like a straw broom, wide and deep. It allows the fish to maneuver in tight spots and provides the surface area for quick bursts of speed. This tail accounts for the ease with which the striper inhabits the surf zone, where the tumbling waves, boulder-strewn shores, and swirling sands send most fish scurrying for calmer seas. It's not the best tail for covering large distances with ease. A better choice would be one that is more deeply forked, like a marlin's or a bluefish's—the forked shape provides less resistance and more speed. But this seems of little bother to striped bass, some of which log more than two thousand migratory miles a year.

The northbound leg begins in the spring when most stripers age two or older drop out of the Chesapeake Bay, where the majority of the migratory stock is spawned and reared. It's estimated that this coursing stream of pinstriped silver is made up of twenty-five million individuals. The spring migration is a sleepy affair, with all fish shaking off the chills of winter and many recouping from the rigors of spawning. The stripers are thin and worn out. The Chesapeake stripers are joined by a relatively small number of fish that spawn in the Delaware Bay and the Hudson River. As they move north, large schools fall out and set up shop in bays and estuaries from New Jersey to the Bay of Fundy. Here the striped bass spend the summer, like parasol-toting vacationers, lolling in pleasant conditions. Then they rally for the fall run.

It starts with aggressive feeding. The weakening angle of the sun's rays, like Pavlov's bell, signals an instinctual urge to eat. Areas that had been close to dormant in July and August boil with fish come September.

Next, the first cold snap of the year barrels down from Canada and sends water temperatures plummeting for the first time in six months. As a result, the trillions of microorganisms riding the tide since they bloomed in spring die off. Almost overnight, the water, which has been clouded by this boon of life, clears up, until one can see a shoal ten feet beneath the surface. And while the bass don't feed on these tiny creatures, their prey does. As the underwater world takes on a new clarity, baitfish—silversides, bay anchovies, sand eels, bunker, herring—understand it's time to go. And as the bait goes, so go the bass.

This domino effect begins in Maine, where autumn gains a stronghold by mid-September. Here the migration starts off with a trickle as thousands of stripers begin to move south. They have the farthest to travel—some one thousand miles—before they reach their wintering grounds. On average, the fish will log about twelve miles a day, but the speed of the migration changes with the weather. Indian summers find the fish dragging their tails. Conversely, a succession of early winter storms chase them south, invoking the common expression, "When the snow flies, the bass go by."

As any college student knows, it's hard to throw a hell-raising party with minimal numbers, and for this reason the fall run in Maine is a shadow of what it will become. Along the way, these fish will pick up other migrants, until there are waves and waves of stripers coursing down the coast. Within a single hour a particular drop goes from barren to chock-full of fins.

The route, while essentially north to south and within a mile from shore, can also vary. Since the task of their migration burns energy, the fish need to eat, and they follow the bait. If the peanut bunker or blueback herring get blown offshore by a west wind, the bass will remain out of reach of surf casters on the mainland but delight those on the islands, like Martha's Vineyard. When the reverse happens, the residents on the Vineyard and other islands can feel as if they're, well, stranded on an island.

The fish in this giant underwater march sort themselves out by size. The two-year-olds—they're roughly fourteen inches long—usually leave first. This will be their first fall migration, and their eagerness to depart is not a result of youthful restlessness but rather a fact of locomotion: Their tails are up to twice as small as their older brethren's, and they need the extra time to cover ground.

It follows, then, that the large cows leave last. (In striper parlance, any fish over fifty pounds is a cow. The largest fish are always females; males rarely reach weights of more than twenty pounds.) Often, these fifty- to seventy-pound behemoths, many of which have been on the migratory circuit for more than twenty years, leave their summer hangouts when fall foliage has become nothing but a brown, brittle ground cover. Among the cows there can often be found a few smaller, blind stripers. And while it would be comforting to anthropomorphize this combination—the matriarchs taking care of their wounded flock—there is no known explanation for this phenomenon.

The touchstone of the fall run is the feeding frenzy known in angling argot as a blitz. During a blitz, baitfish are pushed to the surface by hungry stripers. Jammed against the top of the water column, they form a tight ball that darkens the water, each fish seeking safety among thousands of others. With this wad of baitfish spinning like a disco ball, the first striper launches an attack on its flank. The ball splits wide open. On cue, the other stripers pounce, like children chasing candy from a piñata, and the surface is churned into a mélange of blood and guts. The ruckus attracts terns and gulls, which hover close to the water, waiting for the baitfish to leap from the jaws of the stripers, and swipe their meal from midair, all the while squawking encouragement.

For the angler, hooking a striper in this situation is as simple as landing a lure in the fray. But the blitz can end as suddenly as it began. When it does, fishermen scan the water, waiting for the next round. Below the surface the two sides regroup.

Stripers and bait schools aren't the only animals traversing the rim of the Atlantic in autumn. Terns, gulls, and gannets fly endless patrols searching for blitzing fish. Above them, skeins of sea ducks and geese quack and honk their way south in shaky Vs. The most unassuming traveler of fall, the monarch butterfly, quietly rides the wind with a determination that will bring it to the mountain forests of central Mexico.

But it's beneath the sea's surface where there's an unfathomable torrent of life—everything from sea robins to sharks to stingrays fall in step with the bass. False-albacore schools zip here and there like schoolkids set loose at recess. Blowfish furiously work their stubby tails. Right whales breach just offshore on the way to their breeding grounds off the coast of Georgia. And while not every marine species is headed south—some, like the bluefish, move offshore to wait out the winter at a depth where water temperatures remain static—the entire coastline is quivering with movement. The fall migration is a symphony of life, its music created by billions of fish tails.

3

Maine: On the Road

Labor Day, 2001. As I left New Hampshire, I-95 morphed into the Maine Turnpike. Traffic heading south, just across the grassy divide, was stop-and-go, vacationers leaving summer behind and ushering in autumn in a snarl of cars.

I was whooshing along in the company of a couple of timber trucks. Ensconced in my full-size, 4 × 4 Ford Explorer, I felt like I belonged. The truck wasn't really mine; it had been graciously loaned to me for three months. It was more posh than most fishmobiles, but I had been told to make it my own. The back was loaded with tackle, including eleven rods and reels. There were also duffel bags full of clothes and foul-weather gear, two pairs of waders, camping supplies, and all manner of atlases and gazetteers. A maroon sea kayak, which had already lost its jury-rigged cockpit cover (a Hefty bag secured with bungee cords) in a freak squall on New York's Throgs Neck Bridge, was strapped to the roof.

When I turned off near Bath, Maine, it was dark. I followed the promise of a blue road sign that advertised lodging. Bath proper was still a few miles ahead, but I couldn't afford the motel rates in the center of town. After nine hours on the road, I wasn't picky. I found a small, run-down inn. The front-desk clerk, who was also the waitress, was about to call it a night.

"Any vacancy?" I asked sheepishly, as there wasn't a car in the parking lot, and the dining area reminded me of a dollhouse, its empty tables set waiting for imaginary guests. "There's plenty of room at the inn," she said, quite pleased with her answer. "Water or road views?" I chose the road, since the rate just squeaked in under my $35 lodging allowance and I could keep an eye on the truck.

When I opened the door to my room, a roach made a break for cover under the bed. A twelve-inch black-and-white TV flickered in the corner. I could hear a steady *ping* of water dripping in the aluminum shower stall, and the rumpled brown bedspread sported a monumental stain that bore an uncanny resemblance to South America. I called down to the front desk to see about upgrading. No answer.

Before going to sleep—in my clothes on top of the sheets—I found an outlet, powered up my laptop computer, and wrote my first journal entry of the trip. It began, "Road Rule 1: Don't find a place to stay in the dark."

||| ||| |||

My decision to settle on Bath as the northern terminus of my journey was more enlightened. The quaint seafaring town sits snug on the most classic striper water in Maine, the Kennebec River. While some migrating stripers can be caught as far north as the Bay of Fundy, the majority of the fish that reach the state find the Kennebec and the coast-line that flanks it irresistible. Here, two of the cornerstones of striped-bass fishing, viable habitat and bait, occur with such frequency that the only problem for the angler is deciding where to fish.

Nine native sea-run species, including a handful of stripers, use the river as spawning grounds. They reproduce beyond the tides' influence, where the water runs sweet. Many of the young fry drop downriver to Merrymeeting Bay, a huge, fertile, shallow body of water referred to as the "Chesapeake of the North," to mature. Here, six other rivers, which drain one-third of the state, meet in the nine-thousand-acre brackish estuary. It is a fish nursery non-pareil. For a century or more, however, it was a death trap.

When the industrial age staked its claim on the Kennebec, the river became choked with pollution from timber mills. In 1837 the construction of the Edwards Dam in Augusta, a 25-foot-high, 917-foot-wide monstrosity, squelched the flow of water into Merrymeeting Bay. The dam powered saw mills and a burgeoning textile industry. Sitting just above the tidal reach, it also stymied every sea-run fish looking to spawn. By the 1950s, the waterway's foul stench was forcing employees in the Maine State House to shutter their windows during the summer. Parents forbade children from playing on the river's shores because it was impossible to remove the noxious odor from their clothes. Instead of fixing the problem, man left the fallow Kennebec and began harvesting the sea's crops, where the plumes of pollution had yet to suffocate the wildlife in the vast waters.

Today the river has regained much of its youthful complexion, a testament to the stubbornness of Mother Nature. The pollution was curtailed after the Clean Water Act of 1972. And in 1999 a grassroots effort to rid the Kennebec of the Edwards Dam proved successful, freeing up seventeen miles of river. It was a major victory for environmentalists, anglers, and anyone captivated by flowing water. The day after the dam was destroyed, there were reports of fish nosing into the haunts of their ancestors. Biologists predicted that the removal would eventually double the river's striper population, not to mention boosting stocks of the blueback herring, shad, and alewife, all important forage species.

But the Kennebec wasn't the only reason I started in Bath. I wanted to fish with the most knowledgeable and talented striper guides to get a true sense of the fall run. I had spent the summer E-mailing and phoning guides along my route. Since most charged around $400 a day, I had no chance of paying for their services. I simply told them about my journey, explained my situation, and offered to shell out gas money. I wanted them to treat me like a fishing buddy, not a client. I was pleasantly surprised by their benevolence. The first call I made was to Captain Dave Pecci, a fly-fishing guide who docks his boat a few steps from downtown Bath. Numerous sources told me Pecci was one of the best guides on the Kennebec. When I reached him on the phone, he said he'd be more than happy to show me his river.

III   III   III

I had allotted nearly two weeks for Maine, longer than any other state on my itinerary. An early-season cold front would send the stripers on their journey, and I didn't want to miss any action. On the other hand, an Indian summer would do little to induce the instinctual urge to migrate.

Rather than gamble on the weather, I decided to hedge my bets.

On my first day, I arrived at Reid State Park, just north of the Kennebec's mouth, eager to see swaths of feeding fish, but the Gulf of Maine lay before me as flat as a bowling lane. It shimmered under the noon sun, which was pushing air temperatures into the mid-eighties. Instead of fishermen lined up along the sand, I found a trio of college girls basking in bikinis and two kids building a sandcastle while their mom read a tattered copy of *Bridges of Madison County*.

Undaunted, I pulled on my waders, chose a rod from the back of the truck, and slung my surf bag over my shoulder. As I walked south down Half Mile Beach to the mouth of the Little River, I spotted a jumping sturgeon flexed like an armored apostrophe above the gulf. The sturgeon is essentially a living fossil, a relic from the shallow seas that covered the earth 265 million years ago. As if rejoicing in their species' longevity, sturgeon, some of them four feet long, leap skyward, their white bellies glistening like polished ivory in the sun.

Scientists don't know why these fish have a penchant for free jumping. Some think it's a social behavior, not unlike yelling, "Yo, I'm out in the toolshed" to a friend rapping on the door of your house. Others think it's an attempt to rid their thick hide of parasites. For the rest of the afternoon, the silence of the fading light was broken by the splash-landing of sturgeon.

I had come to Reid looking to land the first striper of my trip. With the tide ripping out of the Little River into the gulf, I was in a likely spot. As the currents swirled and eddied past my legs, I knew that baitfish, who once caught in the flow could not swim against it, were getting swept into the gulf. In these conditions, stripers weave in and out of the current picking off delectables. But after dark, with only a few tinker mackerel to my credit, I left. My hopes now rested with Pecci. I planned to meet him the following day at 6:00 A.M.

III   III   III

At 5:45 A.M. it was dark and, save for the soughing of the river, eerily quiet at the Kennebec Tavern and Marina. The headlights of Pecci's truck cut through the darkness and we were soon stringing our fly rods aboard his twenty-two-foot cuddy cabin. After idling out of the marina, the electronics glowing an eerie yellow, Pecci opened up the throttle and the boat rose to an easy plane. We zoomed under the Sagadahoc Bridge, the sound of the engine echoing off the concrete. On our starboard side, a massive blue dry dock— 750 feet long, 180 feet wide, and 77 feet high—was lashed to shore. It belonged to the Bath Iron Works, a shipbuilding operation that had been the anchor of the town's economy for 117 years. I had driven by the company's main entrance- way the previous morning. Workers with metal lunch pails filed beneath a large sign that proclaimed, THROUGH THESE GATES PASS THE BEST SHIPBUILDERS IN THE WORLD.

"They had that dry dock built in China to save money, and then it was towed fifteen thousand miles across the ocean," said Pecci. The journey took seven months, partly because the dry dock was blown off course by a typhoon. "They've had it since February, and it still doesn't work right." Around the dock, workers had begun calling it the "Yugo of the Seas." Like many in town, Pecci balked at the eyesore nestled on the riverbank.

The dry dock signaled the end of an era. For years, ships of all sorts, from square-rigged cargo vessels known as Down-Easters to navy ships, were launched by sliding them down wax-and-grease-coated ramps called "inclined ways." Launchings were ripe with anxiety, and often the entire town showed up to watch a ship hit the water. In one incident, a nut that fell on a greased way stopped a trawler cold.

Pecci has been launching his own boat into the Kennebec since he was twelve years old. He started guiding in 1990. "It was a way for me to justify spending as much time on the river during the summer as I wanted. That, and being in construction for over twenty years was starting to hurt too much," said Pecci.

When the sun rose, the day quickly took shape, a carbon copy of the previous, crafted by a high-pressure front stalled over the state. The forecast was calling for a high near eighty, light winds, and clear skies. As we drifted over sandy flats, casting "grocery flies" (so named because of the long grocery list of materials needed to tie them) with our fly rods, an osprey sized up breakfast opportunities from above. The river twinkled with baitfish.

"She looks like she's ready to bust wide open," said Pecci. There were massive schools of blueback herring and ale-wives dimpling the surface, and occasional batches of sand eels, wiggling like pennants, scooted beneath the boat. A brilliant day, yes, but the stripers were nowhere in sight. "The high pressure," said Pecci, "always keeps stripers out of the shallows."

The way I understand it, a high-pressure front affects fish the way a hangover does humans. The escalating barometer causes them to cast aside ambitions in favor of lounging in deep, cool waters where the effects of changing pressure are lessened and the bright sun can't reach their sensitive eyes (they have no eyelids). Fish may not need a bicycle, as Gloria Steinem once wrote, but if they had sunglasses they would wear them during periods of high pressure.

We tried our luck in a small channel that drained a cove not far from Fort Popham. This semicircular granite structure was built during the Civil War and later garrisoned for the Spanish-American War and World War I but was never engaged. We had little action in the cove ourselves and also moved on. Fearing another fishless day creeping up on me, I asked Pecci about the fall run in Maine.

"It's not as long, big, or consistent as you get farther

down the coast, but it happens, all right. It's not uncommon to have three or four phenomenal days and nights of schools and schools of fish stacked up. And then they're gone. You don't know when it's going to happen. Or how long it's going to last. But it's pretty great when it does. As soon as this weather turns, things should break out."

Outside of the river's mouth, we targeted points where we had a chance of getting our flies down deep near some rocky ledges. Near Small Point, Pecci told me about a fall night of excess that still haunts him.

"I came here with a friend back in the seventies. It was an overcast night, and there wasn't much you could see. But real big bass had bunched up in this hole real tight. They were getting ready to leave. All we had to do was cast out a live eel, count to ten, and then set the hook. Most of the fish were thirty pounds or better. We could've sunk the boat. We sold them for $1.25 a pound. Back then these fish seemed like a never-ending resource. Nowadays, I'd buy back every one of those stripers for three dollars a pound if I could."

Pecci's story is familiar to anyone who fished hard for striped bass in the late 1970s. The ocean seemed packed to the gills with cows, and with the lenient regulations, it was common practice for anglers to sell their catch. A good night of fishing could net a few hundred bucks. But large fish form the backbone of the breeding stock and as they were weeded out so were their contributions of billions of eggs each spring. Ironically many former "meat fishermen," after seeing the effect of this carnage, became strict conservationists. Pecci is a fine example.

He fought hard to raze the Edwards Dam and serves on the Maine Recreational Fishery Advisory Council and the Atlantic State Marine Fisheries Commission's Striped Bass Advisory Panel. He also advises his clients on catch and release. "I tell them that we worked our butts off to get these fish back, but if anyone wants to keep one, please do. Just be sure you're going to eat it and not give it freezer burn." Apparently his clients take heart: Out of roughly seventy trips a season, Pecci throws only about fifteen stripers into his fish box. "I like to say I've evolved," he said.

As we neared a crevice between two garage-size boulders, Pecci motioned for me to make a cast. "There should be a fish there," he said. I let the fly sink, and with the first strip a striper latched on to it. The pull of the fish felt like a jolt of electric current coursing through the line, and the thin rod danced about. As stripers go it wasn't big, just eighteen inches, but it was the first of my journey.

"That's Fred," said Pecci. "Glad he was home."

III   III   III

The next morning I didn't go looking for striped bass, exactly. I went looking for a man who more than once claimed he'd been a striper in a previous existence. How else, John N. Cole would ask, could one explain his total obsession with the fish.

In 1978 Cole penned literature's most endearing paean to the striped bass, *Striper*. The book chronicled Cole's pursuit of the eponymous fish as a bayman on Long Island's East End some twenty-five years earlier. Cole made no apologies for commercially harvesting his "oldest and most important friend." The experience had brought the two together.

Cole did finger the polluting of the striper's spawning grounds in the Chesapeake and Hudson as a cause for the fish's decline in the years preceding his book's publication, and he made a passionate plea for change. "Consider then the dark forces it would take to extinguish the striper's fires, to end the survival of this most determined survivor. What would be the message for the men who have assigned to this fish the qualities of courage, the virtues of bravery and strength? If this fish were to vanish, how much time would be left to the men who extol it," he wrote.

The striped bass was going the way of the passenger pigeon. Cole was determined not to let it happen on his watch.

The book was read by avid anglers and unabashed landlubbers; one of the latter was Senator John Chafee of Rhode Island. He finished it in an afternoon and the next day began pursuing the necessary legwork to include the striper in the Anadromous Fish Conservation Act of 1979. (Anadromous fish can swim freely from salt- to freshwater. Stripers spawn in freshwater.) This eventually led to the Atlantic Striped Bass Conservation Act of 1984. I, like every striper fisherman on the East Coast, owed one to Cole for saving my fish.

I found his number in the phone book and called requesting an interview. "Of course," he said. "Come by tomorrow around one. The Sox are playing at noon, but I've given up on them for the year." He gave me directions to his house. "You'll have no problem finding it," he said. "Just follow your nose."

Heeding the axiom that the gears of life are lubricated with alcohol, I arrived at Cole's house in Brunswick with a fine bottle of scotch. His wife, Jean, who was working in the garden beneath a straw hat, directed me to the office door. Before I could knock, it swung open. Cole, dressed in a button-down shirt, khakis, and a pair of loafers, darkened the doorway with his tall frame. He looked me over and in a booming voice said, "You're young."

"He tells that to everyone," Jean hollered from a row of tomatoes. "It's because he's so damn old." Cole, seventy-eight, responded with a tremendous laugh, patted me on the shoulder, and welcomed me to his home.

On a tour of his small basement office, Cole told me he

"wasn't horny for fishing anymore," but the place was littered with angling paraphernalia, photos, and books. He peppered me with questions about my luck in Maine, and when I told him of my plan to follow the stripers, he couldn't contain his enthusiasm.

"I love the fall run," he said. "It's an incredible sense of activity, especially farther south. You have all those fish, not just the striped bass but bluefish, albacore, bonito, tearing up the place, just gorging themselves. Birds go wild and they scream. It's very high-pitched. It's an incredible sound. It's the only time of the whole twelve months that there is that much energy in one place. It's in the air, on the water, underwater, in your own interior. You could put a librarian who has never fished before in the middle of a blitz and her adrenaline would run."

Cole knows of adrenaline. At nineteen he left Yale to fight in World War II. During his stint, he survived thirty-five combat missions in a B-17 and received numerous medals. "The medals aren't important; the important thing is that I'm here," he said. "Half the guys didn't make it." Back in the States, Cole graduated from Yale and went to work for a public relations firm in New York City, but he felt stifled by city life. "I just said screw it, I'm not doing this anymore. And I went out to the end of Long Island, where I had grown up."

Cole became a member of a dory crew, essentially a gang of five men who rowed wooden boats into the surf and set nets for striped bass. "There were no politics involved, just you and the fish," he said. This type of fishing is rarely lucrative, unless, like Cole, you mine more than monetary rewards from the sea. "Those guys I fished with were the best friends I've ever had," he said. "That experience made me who I am."

He has been a tireless defender of everything wild since he accepted his first newspaper job from Joe Pulitzer, a man he had taken fishing on the East End. In 1968 Cole cofounded the *Maine Times*, an earthy, hard-charging weekly that barked and bit at the heels of any industry polluting or deforesting the state. When the Presumpscot River was rendered foul and stagnant by the S.D. Warren paper company, Cole threatened in an editor's note to pay for a hot-air balloon that would be permanently anchored over a nearby highway. On the balloon he would paint the words "Nobody farted, this smell is brought to you courtesy of S.D. Warren."

"The river belonged to the people," he told me. "I couldn't understand how in the hell they could stand it."

In 1984 Cole worked with his friend Brad Burns to restore a native spawning population of striped bass to the Kennebec. Cole gathered the funding while Burns figured out the logistics. Eventually they had striper fingerlings trucked from New York and released in the river. Some years later their effort was rewarded when twenty-six newborn stripers were netted in Merrymeeting Bay. In a local newspaper story about the event, Cole proclaimed it "the best news of the century."

Cole invited me to stay for lunch, and I gladly accepted.

Jean made BLTs with fresh tomatoes from the garden. In the middle of telling a story about Jean's first trip to Montauk, Cole jumped from his seat, grabbed a BB gun, and fired at a squirrel raiding the bird feeder. "I don't know if you've noticed, but we have a distinct species of squirrel in the yard—we call it the welted squirrel," he said.

On the wall hung a calendar where Cole recorded the wildlife he saw each day. As we ate, his head followed every bird that flitted by the window. I'm sure that after I left, a few passersby made the log.

I had come with one question for Cole. "What is it about the striped bass?" I asked.

"The thing about the striped bass is that they're beautiful. They're also the wildest, biggest fish that's inshore. And they're very loyal. You can walk out to a salt marsh anywhere along the coast and catch one. They show up every year in the same spot, unlike the bluefish, which is evasive and mercurial. You develop an affection for a wild creature that comes to you, that puts itself in your environment. Stripers allow us to connect with the natural world. And they taste pretty damn good."

||| ||| |||

At Reid State Park that afternoon, as if on cue, the stripers displayed their loyalty. Casting a lure painted to resemble a tinker mackerel, I came tight with five nice fish. And over the next week, I found my groove on the road. I studied maps and guidebooks, stopped to chat in every tackle shop I passed, and pulled over to fish any piece of water I fancied. Day after day I watched the sun rise and set from the fishiest water in the state. While the Indian summer delighted locals, I kept an eye on the weather, praying for the cooling winds of autumn to blow. The stripers had yet to show their fall colors, so I concentrated my efforts at peak feeding times, the happy hours of dawn and dusk and throughout the night. At these times, stripers rely on their excellent night vision to ambush unsuspecting prey.

To catch a striper at night, you need to reel agonizingly slowly. Old-timers like to say, "Reel as slow as possible, then reel twice that slow." With a good cast, one retrieve can take at least three minutes, and often my mind wanders from the task at hand. I watch clouds scud across the sky. Count shooting stars. A fish strike yanks me by the arms from slow motion to real time. It can be as startling as someone sneaking up behind you in a quiet room and blowing an air horn.

To get a strike, however, you first have to make a decent cast. Without visual cues, the whipping motion that requires no thought during the day suddenly takes on an awkward life of its own. The first time I ventured into the surf at night, I stepped into the water and without hesitation reared back and let fly. What I didn't know was that my line had wrapped around the tip of the rod in a knot. The force of the cast broke my line. It parted with a *snap* as my best lure went sailing into the suds. Not too long afterward, the spool of my reel somehow worked loose and dropped at my feet. I was in white water up to my knees. I pulled on the line, but the spool simply rolled out to sea. Eventually the line hung up on a snag and I had to cut it. I went home dejected.

Luckily, my gear and senses held together in Maine. One evening around 11:00 at Higgins Beach, a postcard-perfect beach town hidden just south of Portland, a pod of large bass worked close to shore on the heels of a thick fog. I was casting a large swimming plug called a Bomber. It's seven inches long and as big around as a broom handle; three treble hooks the size of an infant's hands hang beneath it. The resistance of the water on the lure's wide plastic "lip" causes it to wiggle seductively when reeled in.

The first striper smacked my Bomber within three feet of my rod tip, opening a hole in the surface as big as if a Virginia ham had fallen from the sky. I lunged backward in surprise, pulling my lure out of the striper's mouth. On my very next cast, I connected with a fish well over thirty-four inches, probably fifteen pounds. "They're here," I said to myself. This was not the fall run, just a typical summer pattern—a school of big stripers feeding close to shore under the safety of darkness. I didn't mind. I was catching fish, and I was completely alone.

The bite stopped at 3:00 A.M. Slack tide. I decided to move up to a dune for a quick nap while the tide turned. Trudging up the beach in the fog, I practically tripped on the wooden ribs of the *Howard W. Middleton*, which rose like a battered fence from the sand. They were all that was left of the grand sailing ship. Bound from Philadelphia to Portland in 1897, the double-deck schooner got lost in what was described as "pea-soup fog" and ran plumb into the beach with a cargo full of coal. When the tide ebbed, the sailors swam to safety, but the ship eventually broke up, spilling 894 tons of coal into the ocean. For weeks, when locals needed fuel for their stoves, they simply wheeled a cart to the water's edge. In the days following the wreck, local boys salvaged the ship's brass bell, but the police chief made them return it to the owner of the ship. I pulled my own souvenir from the timbers: a splinter of wood washed smooth by wave and tide.

With my head resting in the sand, I fell asleep. I woke to a dog sniffing my foot. It was morning. The dog's owner, an older woman in khakis and a sweatshirt, was a few steps away. "Oh goodness, you scared us," she said. "Everything okay?"

"Absolutely perfect," I said. "Except I slept through the dawn bite."

III   III   III

By late afternoon, my nap on the sand and seven straight days of casting had taken their toll. My shoulder was sore and my back ached. I checked into the Howard Johnson's in South Portland, wanting nothing more than to order a pizza and go to sleep. In the morning I would need to find a Laundromat. My clothes had become unbearable. The truck reeked like a bait bucket. It was September 10.

When I went to bed I could scarcely have known that less than a mile away in a Comfort Inn, two men were rehearsing their final plans to fly an airliner into the North Tower of the World Trade Center the following day.

I spent the morning writing and was unaware of the tragedy until noon, when I started the truck. I sat stunned, listening to radio reporters describe the scene. I ran back to my room and frantically called my parents. My brother's investment-banking firm had a few offices in 7 World Trade Center on the thirty-second floor. He was often there for meetings.

My father answered the phone, which is a rarity, and I assumed the worst. He could sense the fear in my voice. "Your brother's fine. He was scheduled on a flight out of LaGuardia at noon and didn't go to the office."

"What's going on?" I asked.

"I just watched the World Trade Center buildings disappear, Dave."

"They're gone?"

"They're gone."

When I finally reached my girlfriend, she asked simply, "Can you come home? It's like a war zone outside. I'm really scared." With the city locked down, getting home was not an immediate option. We made plans for her to visit her parents in New Jersey as soon as the Hudson River crossings opened.

I left for Yarmouth, where I was supposed to meet Pat Keliher at the Coastal Conservation Association (CCA) office. I turned into a small shopping plaza and marina on the banks of the Royal River. From the truck I watched an older man framed against a preternaturally blue sky lower an American flag to half-mast. On the river the tide was slack, and the boats shifted on their moorings like horses hitched to a post.

Too numb for most of the morning to cry, I began to sob when I heard a firsthand report of people jumping from the towers. I walked to the bank of the river, where for the next two hours I debated the fate of my journey.

III   III   III

The following day Pat Keliher and I decided that an afternoon on the water might help us take our minds off the situation. We met directly across from the marina where I had parked the previous day.

"I had to get away from the TV," said Keliher. "I couldn't take much more." Aboard his eighteen-foot Parker, we cruised out to Casco Bay. The sky was empty—not a cloud or a plane. Even the assiduous lobstermen had halted work for the day. We veered around a patch of ground hidden by a few inches of water. Keliher pointed to a heron standing guard. "If I ever write a book on boat handling, I want to call it 'You can't run a boat where birds are standing.'"

Keliher pulled up on the throttle near a large grass flat that was pockmarked with sand patches. I stepped to the bow and worked out a few practice casts with my fly rod. As I loosened up, Keliher finally shed the thick gloves he'd been wearing, even though the temperature was nearing sixty-five degrees. "My fingers get cold fast," he said, grabbing a push pole from the gunnel. One January, while hunting sea ducks, Keliher got a nasty case of frostbite while retrieving a string of decoys. He's reminded of his mistake on all but the warmest days.

For seven years Keliher was one of the hottest striper guides on the coast. On one memorable day, two of his clients boated 840 pounds of striped bass on fly rods. "We released every fish," said Keliher. "Some of them were over forty-five pounds." Keliher pardoned all but three stripers during his guiding days. "I knew those fish would die even if I released them, but it still pained me to put them in the box." In 1996 his conservation ethic and natural leadership skills prompted an offer from the CCA to become Maine's executive director. Stripers are always on his organization's radar.

III   III   III

Sight fishing is a fly angler's ultimate pursuit. It requires patience, stealth, and a boatload of skill. Even Shakespeare gave it a superlative nod in *Much Ado About Nothing:* "The pleasant'st angling is to see the fish / Cut with her golden oars the silver stream / And greedily devour the treacherous bait . . ." Most sight fishing is reserved for the flats of tropical locales, like the Bahamas or the Seychelles, where the wily bonefish and permit swim. In the Northeast, sight fishing is limited to small pockets of the coast, and the striper is the only game in town. Of course, Keliher discovered one hell of a pocket.

"Don't look for fish. You'll never see them," Keliher told me. "Look for their shadows on the sand, then you'll spot them." He was right. Minutes after we started our drift, I noticed a shaded patch of bottom moving against the tide, like a cloud drifting over a ballpark on a sunny day. I spotted the fish, a school of five bass nosing along just above the bottom. I made a false cast and then delivered the fly. It arrived late. The bass weren't spooked, but before I could gather my line they were behind the boat, out of reach. "It's all right. We'll have plenty more shots at them," said Keliher. And we did. The highlight of the day occurred when a twenty-pound striper lumbered into view. With my knees shaking, I lobbed a cast in its general direction, but it swam past as if it couldn't be bothered.

For most of the day I had trouble finding a steady rhythm. My stroke was slow and my casts often landed short of my targets. Finally, I managed to hook a small striper on a sand-eel pattern. As it locked on to my fly, it gave a quick flick of its tail, turned sideways, and engulfed it. The fish's schoolmates scattered, leaving puffs of sand riding the tide. I fought the fish to the boat and Keliher deftly scooped it up. We admired it for a long time. Then he slid the fish back and we called it a day.

Our experiment had worked, but it was time to motor back to reality. As we straightened up the boat, we heard the first plane of the day droning in the distance.

As if shaken loose by the horror of September 11, the high-pressure front lost its grip over the Northeast that night and was replaced by a low-pressure system. Snow was predicted in the inland areas of Maine; along the coast, rain and northwest winds.

III   III   III

The next afternoon when I pulled into York, Maine, I was greeted by a welcome sight: blitzing fish. Just off a stretch of beach that fronts this small town, birds were wheeling and diving above a pocket of water near the Nubble Light jetty. The run was on.

I pulled a quick U-turn, parked the truck, and ran for the beach. "No need to rush," a man yelled from his car. "Those fish aren't going anywhere. And take that popper off. You need to throw metal at these fish. Try a Hopkins or a Kastmaster." He looked at my New York license plate. "They're feeding on what you call peanut bunker," he said. "*We* call them pogies in Maine."

The people of York have kept an eye on visitors since the Candlemas Raid of 1692, when Abenaki Indians, encouraged by the French, walked into town and killed around one hundred residents and carted another eighty back to Canada as slaves. They also burnt forty of the forty-five homes in the village. Legend has it that before the raid, the Indians left their snowshoes in a pile by a large rock outside of town. When a local saw the snowshoes, he ran to warn his fellow villagers but was too late. Though one-third of the population was lost, the town regrouped and moved to higher land.

My de facto guide, it turned out, was absolutely right. The fish stayed tight in a corner, marauding schools of peanuts till the sun went down. Working my Kastmaster, I landed twelve before I lost count. Afterward I walked over and introduced myself.

"Hi. Thanks for the tip," I said, as I extended my hand.

"Robert Magocsi. York, Maine. Pleasure to meet you."

A lifelong resident of York, Magocsi spends a part of every day working the beaches. His neoprene waders bulged a second time above his protruding gut, where he stashed his pack of Merits, a lighter, and tackle in a Zip-loc bag.

Barking instructions on the beach seemed to come naturally to Magocsi, a vestige of his military days as an instructor in an officer training program. "I was on the beach once when George Bush Sr. pulled up in his boat," said Magocsi. "We were hauling in bluefish and he couldn't hook a thing. He came over and asked me what I was using. I told him they were hitting anything you threw at them. I like to say I've been a personal adviser to the president."

Families holding candles lined the beachfront road as I pulled away, the joy of the afternoon quickly tempered by reality.

The next day I met up with Dave Gittins, a fly-fishing guide who keeps his boat on the York River. He was jazzed about our chances. The previous day, Gittins had run into a school of fish just minutes from his dock. The action had been so good that he radioed a fellow guide to let him in on the frenzy. They spent the day there. "It's happening," he told me.

We turned south after leaving the river and cruised down the coast. Within minutes we were into fish. Near the shore, pods of stripers balled up a school of peanut bunker and thrashed at them. Gittins motored in and we launched our flies. Soon we were whooping and laughing. The action got so heavy that Gittins's rod snapped in two.

With the migration under way, my trip was picking up momentum. The fish were starting a journey that would carry me along like a piece of flotsam on the tide. They were following the sea road now.

4

New Hampshire:
Running and Gunning

New Hampshire, like a bathing beauty, dips only a toe in the North Atlantic. As a result, striped-bass aficionados often overlook the state's twelve miles of coastline. Even resident anglers prefer to think of themselves as perfectly situated for expeditions to Maine and Massachusetts.

I wasn't due in Boston for a week, so I had no choice but to cover every inch of New Hampshire's coast. The shoreline was not furrowed, like Maine's, but more like a blunt edge. It was easy to see why Samuel de Champlain, a meticulous note-taker, made few remarks about this stretch of sand when he explored the area in 1606. Migrating striped bass also have been known to bypass the state completely. It happens when the bait schools get blown offshore toward the Isles of Shoals, six miles east of the coast. The stripers follow their food cart, reuniting with land when they brush against Cape Cod. Thankfully, this wasn't one of those years.

The method of choice for many surf fishermen is called the "run and gun." You hop in your car and drive down Route 1A, which snakes just beyond the beach, glassing the water with binoculars at paved turnoffs. If action beckons, you grab a rod and run; if not, you keep moving. I found myself following a circuit that started at Odiorne Point State Park, on the outskirts of Portsmouth, and finished in Hampton. It's not too scientific, but with the migration in swing, fishing one spot can be a boom-or-bust situation. When rumors of fish spread, trucks and cars with eleven-foot surf rods sticking out of passenger-side windows make a bee-line for the action. The week before my arrival, the stripers were feeding close to the beach and anglers were thrilled.

On my first day, the fish moved just out of casting range. There's nothing quite as frustrating for a surf fisherman as standing on a beach and watching stripers blitzing ten yards beyond his plug's splashdown point. If there were such a thing as a lure-launching gun, I'm sure the beach would sound like a skeet range.

As the tide dropped, I waded in and hooked a few
schoolies from some small rock islands with a pencil popper, a lure that wags across the surface like the hand of a
metronome. The water was perfectly clear and just above my
waist. As I explored a sand flat, a pod of peanut bunker more
than forty yards long and about five yards wide approached.
I stood as still as a heron. The school enveloped me like
poured silver. They were thick enough, it seemed, to sit on
and ride. Among the shallows they had found a safe haven
and didn't look in a hurry to leave. I wondered if they could
hear their brethren being ambushed offshore. Eventually
the last peanut bunker slipped by. There were no stripers
tagging behind.

My invitation to the party offshore sat on my truck's roof.
But by the time I decided to launch my kayak, a groundswell
sent by a distant hurricane had begun lumbering toward
the beach. The waves seemed lazy as they approached, but
they crashed with a heavy thud, like a Joe Frazier hook. It
was too early in my trip to splinter my kayak—or my
bones—on shore.

III   III   III

The following morning before sunrise, I spotted a pickup
truck with a Massachusetts license plate parked by a beach
access. Its back window was plastered with an array of
stickers, the largest of which was a red heart with the words

"I love Hooters." From the truck's bed, a bare-chested man pulled out a four-foot wooden speargun. It looked like a giant version of the rubber-band guns I used to hunt down houseflies with when I was a kid. I pulled off the road and introduced myself.

It was chilly out, around forty degrees. A woman on her morning walk passed by, wearing gloves and a wool hat. But Jan Polex seemed impervious to the cold. His wet suit was rolled down to his waist, and the wind was ruffling the hair on his chest. He'd left Dracut, Massachusetts, that morning at 3:30 to be in the water at sunrise. He'd be back in his office, where he develops laser technology for the medicine and aerospace industries, by 7:30, before most of his coworkers showed their faces. The waters in his home state are loaded with stripers, but spearfishing is illegal there. He'd been drawn here by the potential for real big bass. "This is the time of year for them," he said. "Often I'll see the cows following a school of smaller fish. They let the little guys go ahead and check things out first. They're not dumb."

Polex told me he'd stumbled on the sport just two years before. "I was at a dive show and I saw some camouflaged wet suits. They had a rocky-seafloor pattern. I asked about them, and the guy told me you wear them to hunt stripers. I'd been scuba diving for twenty-five years in New England and I'd never seen a striper. I thought he was full of crap.

Then he tells me stripers get spooked by the air bubbles from scuba tanks and to hunt them you need to hold your breath."

Polex purchased a wet suit and a speargun. He practiced holding his breath during his work commute. And on his first foray without a tank strapped to his back, he saw stripers. Now he spends up to four days a week in the ocean.

Spearfishing is a growing trend. The sport first saw its popularity rise when *Sea Hunt*, with Lloyd Bridges, aired in 1957. It attracted macho types bent on slaying leviathans, but remained a fringe sport until the late 1980s, when a new pied piper surfaced in a set of fins and a mask. His name was Terry Maas. A dentist from California, Maas saw the sport as a spiritual pursuit, and he took it to new levels. His targets included giant pelagic species, like three-hundred-pound tuna and marlin. He called it "blue-water hunting" and published books about his adventures. In no time, the sound of spearfishermen plunging off of boats could be heard on both coasts. In the Northeast, the striped bass became the quarry of choice because of its size and its penchant for near-shore environs.

Maas preached safety in the water. That's because the sport is more extreme than many of those favored by Mohawked teenagers on ESPN's X Games. Without air tanks, you must rely on the capacity of your lungs. On land, the average spearfisherman, with a few weeks of practice, can hold his breath for three minutes. But underwater the exertion of swimming cuts that time in half. Pushing yourself can have dire consequences. Shallow-water blackout, a sudden loss of consciousness caused by oxygen starvation, thins the ranks of spearfishermen every year. According to Maas, "The blackout occurs quickly, insidiously, and without warning. Mercifully, the victims of this condition die without any idea of their impending death." Beyond shallow-water blackout, there are a number of variables that can take a life, including hungry sharks. There are also plenty of niggling discomforts, like cold water and even seasickness. In the surging waters near shore, even the best spearfishermen often feel their stomachs flop. To cope with this, they dine on yogurt before splashing down. "It goes in easy, comes up easy," one diehard told me.

When I mentioned the dangers to Polex, he shrugged them off. He was studying the waves. His favorite spot to hunt was about 250 yards out. There, twenty feet down, a rock ledge dropped off into deeper water. Spearfishermen, like deer hunters, bag their prey by ambushing it, and the ledge acted as a game trail. Polex liked to hide behind a large boulder that rests on the ledge. "Big stripers love it. They swim right by me. It never fails," he said.

In position, Polex lays prone on the bottom, head slightly cocked toward the seafloor, and waits for a fish to come into view. "If they see your eyes, they're gone," he said. Many spearfishermen also believe stripers can sense the vibrations from an inexperienced or excited diver's racing heart, causing them to flee.

When the water is clear, and your gun has the firepower, you can shoot stripers from up to twenty feet away, but often the water is murky and you don't see your prey until it's three feet from your mask. Polex experimented with a few different guns before making his own from mahogany. Above the surface it weighs ten pounds, but underwater it's neutral. Two bands of surgical tubing called rubbers power the gun. The rubbers, which each pack one hundred pounds of thrust, fire a stainless-steel spear. If it's a good shot, the spear will break the striper's backbone; in diving parlance, this is called "stoning a fish." If not, the striper thrashes for a few seconds before it can be subdued, usually by pushing a knife into its skull. "I rarely miss," said Polex. "I stone the majority of my fish."

Spearfishermen can't exactly pull out a tape measure before they shoot, so they're extremely careful about killing undersized stripers—and a bit defensive. "Unlike anglers, we get to choose our fish. Fishermen practice catch and release. We practice release and catch," said Polex. His biggest "catch" this fall, which he stoned just behind the pectoral fin, was a thirty-nine-inch bass.

Under the surface, Polex is privy to a scene that anglers can only imagine. Along with stripers, Polex sees cod, pollack, bergalls, tinker mackerel, and bunker, as well as eels, which ride out the daylight hours suspended in the kelp beds. One morning, after the water instantly went from clouded with stripers to barren, he surfaced and saw three seals staring right at him. "They looked like beach balls with whiskers. I think they assumed I was one of them." Polex also salvages enough lures hung up on the rocks and kelp to start a used-tackle company.

With the hurricane still churning up the ocean, large rollers pounded the beach. Rough water increases turbidity and makes spearfishing difficult, if not impossible. Conditions were worse than Polex had imagined, but he was willing to give it a shot. "Sometimes the viz improves as you go out farther," he said. On the beach, he timed the waves and then paddled off. I watched Polex's legs pumping on the surface as he bobbed on the swells. Then his head dipped below the surface, his feet rose into the air, and he sank below the water, like an arrow dropped from the sky.

"It's real quiet underwater," Polex told me before he pushed off. "Sometimes you hear a lobster boat in the distance or the clinking of the pebbles on the shore when the waves are up. It's peaceful. But on some days, in the back of your mind you start hearing the *Jaws* theme. It's amazing what that will do to your breath hold."

Encouraged by Polex's tales of the underwater world, I went to a nearby dive shop and purchased a wet suit. I would need it in Rhode Island, where I planned to do some free diving with a striped-bass videographer, and for skishing in Montauk. I bought a full-body suit along with hood, gloves, and flippers. And even though I held off on the weight belt, I donned my new getup that afternoon and went for a swim. Without the added weight, I bobbed around, my arms and legs higher than my midsection. I felt like a beetle on its back. Not surprisingly, my excursion didn't yield any sightings aside from numerous snails and a green crab.

III   III   III

With the fish out of reach, I was appreciative of an invite to CCA–New Hampshire's meeting to discuss an upcoming amendment to the striped-bass management policy. At stake was everything from size limits to commercial quotas to habitat restoration. It was hard to meet a fisherman who didn't have a gripe about the situation. Fly fishermen, who make up the vast majority of the CCA, felt that lenient regulations allowed too many big fish to be killed—they wanted more shots at cows. Normally the guys with the deepest pockets, fly fishermen rarely keep a striper, bringing over an aesthetic from their freshwater brethren. To them, releasing a fish consummates victory.

Surf casters, mainly blue-collar guys, have no qualms

about dispatching a fish for the dinner table. Most of their beach buggies have giant coolers bracketed to the front bumper for holding their catch. Still, they had seen the results of overharvesting and were happy to pass up a few fish to ensure a viable fishery for their children.

Charter-boat captains in many states can keep up to two fish per client and many would love to get a bigger piece of the pie. Piling fish on the dock like cordwood serves as an effective lure for hooking prospective clients strolling around the marina.

Beyond the recreational sect, commercial fishermen wanted their quotas raised. They were tired of being labeled as the scourge of the seas, especially when they were allowed to kill fewer bass than the recreational guys. In 2000 commercial landings topped six million pounds; the rod-and-reel gang took home seventeen million pounds. The commercial fishermen, however, are happy with the smaller fish, which normally bring more money per pound at market, as they're easier to prepare when cooking.

The brouhaha over striped bass is not a new one. Besides possibly the cod, no other fish has figured so prominently in our country's heritage.

When the early colonists first arrived at Jamestown, they reveled in the abundance and culinary quality of the striper. Captain John Smith wrote, not long after landing,

"The basse is an excellent fish. . . . They are so large, the head of one will give a good eater a dinner and for the daintinesse of diet they excel Marybones of Beefe."

In 1635 William Wood bragged about landing up to three thousand bass in one net strung across a small creek. The colonists knew this mauling couldn't last forever. Four years later, the Massachusetts Bay Colony prohibited the use of striped bass and cod as fertilizer. It was the country's first conservation measure. In 1670 the Plymouth Colony imposed a tax on the sale of stripers, and used the money to fund the first public school. In his book *The Striped Bass Chronicles*, George Reiger writes, "Had our founding fathers chosen a fish rather than a bird as our national emblem, it would have had to have been the striped bass."

Stripers spawned and thrived in every estuary from North Carolina to Canada until the Industrial Revolution, when dams and pollution finally managed to snuff all but the Chesapeake Bay and Hudson River stocks. Coastal anglers promptly turned their attention to bluefish, which spawned at sea. In 1934, however, spawning striped bass in the Chesapeake staged a rally that would send the fish back into the country's collective consciousness. Thanks to optimal conditions, the majority of the eggs matured into juveniles. Two years later, when the bass left the Chesapeake in the spring, fishermen in the Northeast hooted with delight—the fish they cherished was back. But not for long.

With few regulations in place, recreational and commercial fishermen decimated the 1934 year-class. Still, the brief dalliance had revived enthusiasm. Studies were funded to determine the cause of the striper's decline. The government also demanded that scientists develop a course of action to bring the fish out of the proverbial woods. Regulations were recommended for coastal states.

In 1940 another banner year-class was born in the Chesapeake, and when the fish readied to leave the bay two years later, the Atlantic States Marine Fisheries Commission was formed. The first bureaucratic body of its kind, it would oversee each state's policy on striped bass. Things looked rosy, especially when the veterans of World War II returned to the ocean to find solace and discovered a plethora of stripers. It stayed that way until the late 1970s, but there were ominous signs of a coming collapse.

While the ocean was loaded with big breeders, there were no small fish to be found. For some time catching a forty- or fifty-pound striper was more common than landing a schoolie. The stripers were spawning, but their eggs were being poisoned by pollution in both the Chesapeake and the Hudson. As the large fish were caught there were no other stripers replacing them.

By 1982 commercial landings had fallen by 98 percent from a high set in 1972. Recreational anglers fishing an entire season were happy just to get a strike. Actually catching a striper was cause for celebration. The howling of sportfishermen, combined with the popularity of John Cole's *Striper*, spurred legislation, and the Atlantic States Marine Fisheries Commission, backed by Congress, ordered a moratorium in 1984. There would be no more commercial netting and strict regulations were placed on recreational angling. The fish responded. The ban was lifted in 1989, and the stock was pronounced healthy in 1996. The striped bass had become the golden child of fisheries management. By 2000 the population was numbering near forty-five million fish. In the eyes of many fisheries officials, the current debate was just fine-tuning.

The CCA meeting I was attending was held at the Urban Forestry Center in Portsmouth. The walls were decorated with tree-identification tips and overviews of the dangers of acid rain. Many of the participants had come straight from the water. Pat Keliher and Dave Gittins drove down from Maine. With their deck shoes, their weathered faces, and their sunglasses hanging from their necks, they created an interesting juxtaposition to the decor, like a submarine sitting smack in the middle of the woods. There was a small raffle for some lures and fly boxes. A man everyone called Swannee collected the money. He looked sturdy enough to splay the spindly legs of his aluminum chair, and his arms were covered in tattoos. When the number on one of my tickets was called, I approached the table and chose an

Atom popping plug. My choice garnered a few approving grunts and a smattering of applause. Before the meeting was called to order, I chose a seat not far from Swannee. For the rest of the night I could hear him working to get each breath.

One of the three guest speakers had failed to show. Dick Brame, a marine biologist who works for CCA, had been caught in a web of post–September 11 airline cancellations. Gary Shepherd drove up from Woods Hole, where he serves as a research-fishery biologist for the National Marine Fisheries Service. He spends the majority of his time crunching numbers. He wore Top-Siders, a pair of khakis, and a striped button-down shirt. Doug Grout, the other speaker, worked for the state's Fish and Game department. With his bushy beard he seemed one of the few guys in the room comfortable with his woodsy surroundings.

For a long time Shepherd and Grout tried to reduce fisheries science to layman's terms with a series of graphs, charts, and definitions displayed on an overhead projector. It was a tedious endeavor. Trying to follow along, I felt as if I were back in precalculus class.

Eventually the floor opened for questions. "What do we need to do to get bigger fish?" The answer was maddeningly simple. "Just wait," said Shepherd. "The majority of the fish in the stock are eight to eleven years old, and in a few years they should be thirty-six inches or better. This is a healthy stock. With smart management those fish won't disappear."

"If the stock is so healthy," asked a man who was sitting

with his wife, "then why aren't there any big fish around? It seems only natural that there should be older, larger fish." The crowd shifted in their seats, eager for the answer. Shepherd had heard it before. "You have to ask yourself what's healthy. Most commercial guys want lots of small fish. Is that healthy? Many recreational guys want lots of big fish. Is that healthy? As soon as you start fishing a population, you impose mortality and that changes the structure of the stock."

The father of conservation, Aldo Leopold, bumped his head against this dilemma more than fifty years ago. In *A Sand County Almanac*, he wrote, "But all conservation of wildness is self-defeating, for to cherish we must see and fondle, and when enough have seen and fondled, there is no wildness left to cherish."

III   III   III

After the meeting, I swapped fishing stories with a few guys and headed to my motel. I turned down an invite to fish that night since I had an early trip planned for the morning. Instead, I patched two holes in my waders and glued a new tip-top guide on my fly rod. Life on the road was taking its toll on my equipment.

The next morning, my cell phone rang around 5:30. I knew it was bad news. From my motel just off the beach, I'd listened to the wind rattle a loose gutter all night. I picked up the phone, skipping straight to the pertinent question. "What's the word?" I asked.

"She's humming out there. We'd just get banged around and beat up. Better off staying at the dock, I'm afraid," said my guide for the morning.

Another day of running and gunning. I stepped outside. The air smelled like a sand dollar. The wind was whistling out of the northwest at twenty-five knots, sending low, gray clouds skittering over the treetops. As I drove, occasional waves crashed on the shore, splashing water onto the roadway. I recognized a few other cars. We were all scoping out the scene.

Finally I spotted some terns hovering over the water and pulled off. I didn't see any fish breaking, but I was encouraged by the bird activity. My first cast dredged up a brown weed that resembled moss. Mung. It gathered in a loose clump on the lead head of my bucktail jig. On the shore it had collected in a soupy mess. When the sun started to work on it, the smell would be unbearable. Mung is not true seaweed but a member of the algae family. When Thoreau visited Cape Cod in the 1800s, he called the stuff "monkey hair." It's also known as angel hair and brown wool.

While the stench may offend those walking on the beach, the weed is at its worst in its natural element. Mung can form massive islands that drift near shore, fouling anything that comes in contact with it, especially lures loaded with hooks. The algae sticks to lines and tackle like cotton candy, a brown mass of cotton candy. And when you're counting on your lure to entice an instinctual attack from a fish, a glob of weed doesn't enhance its chances. Even commercial fishermen feel the wrath of the nefarious weed. Masses of mung have been known to topple giant fish traps, called pound nets, which are anchored by pilings driven into the bottom.

I've heard that divers often report large stripers hanging just below the mung islands. The problem is reaching them. As always, I tried, but a dozen casts later, with a pile of the discarded weed at my feet, I gave up.

The mung problem seems to have gotten worse in the past century, and some scientists speculate that it has to do with the runoff of nitrogen, which acts as a fertilizer. It doesn't help that the algae is like Freddy Krueger—it just won't die. When it breaks apart it simply regenerates, making more of a mess. Birds, however, have benefited from the mung explosion. They plunder the mats of drying weed for insects, crabs, and other animals.

Mung is not the only "weed" that plagues fishermen. Rockweed and bladder wrack cling to the rocks that litter the tidal zones. The two weeds sport air bladders that keep them buoyant, ensuring that their leaves will receive plenty of sunlight for photosynthesis. When swells wash over these plants, their long stems wave to and fro like so many human arms at a rock concert. They also provide plenty of surface area for lures to hang up on. Ripping your line free often requires a series of fast, forceful jerks with your fishing rod, the marine equivalent of yanking on the pull cord of a stubborn lawnmower. It is less of a problem at high tide, when your lure rides above the mass of foliage.

The insidious oyster thief, a nonnative species that may have found its way here from Japan in a ship's bilge water, doesn't foul lines. It attaches itself to oysters and other shellfish in their formative stages. Later, when the plant has matured, it gets carried away by the current, taking the oyster with it. In his book *Blues*, John Hersey called the oyster thief "the Honda of the sea." A colony of them can wipe out a shellfish bed in a season, removing a prime source of food and habitat for the striper.

The leaves of all weeds do provide a place for smaller creatures to hide out. My favorite, the lumpfish, gains purchase on the seaweed by using its suction cup–like pectoral fins. A poor swimmer, the fish hunkers down to keep from getting swept away by swift currents. This weedy home also offers perfect cover for picking off delectable sea worms and crustaceans.

And while casting into a sea of weeds can be a frustrating experience, I prefer it to the synthetic flora produced by humans. In my home waters of Long Island Sound, where sea lettuce is the most prevalent weed, I have, on two occasions, hauled in a condom. Old-timers call this catch a

"Coney Island whitefish." It's nice to think these remnants of copulation indicate that we have a grasp of safe sex and population control, but I doubt a lumpfish wants anything to do with a used condom. And neither do I.

III   III   III

On my last day in New Hampshire, the wind and hurricane swells finally let up, leaving ample opportunity for the fog to take over. I was meeting Captain Dave Burkland at 6:00 A.M. for a trip on the Piscataqua River. At the dock, there was about twenty feet of visibility, but I headed in the direction of an idling engine—it was bolted to Burkland's nineteen-foot Eastern.

I worried that the fog would be a hindrance, but Burkland was stoked. "I love foggy mornings. Keeps all the riffraff off the river. We'll have the best spots to ourselves if it hangs in here," he said.

At his day job, Burkland is a foreman for a construction company that builds bridges. His no-bullshit demeanor suggests he's a natural for the position. A self-described "product of the California sixties," Burkland's lifestyle changed abruptly when he took a hitch with the army. He served two tours in Vietnam before coming home.

When he's not building bridges, Burkland runs a charter business. He suffers from one minor problem: He lives to fish. Sitting in the stern when the stripers are rolling and watching a neophyte hook himself with a fly rod pains him. With the end of the season only days away, Burkland wanted to make up for lost time.

Lately the bass had been rising from the main channel with the tide and pushing onto the shelflike flats that line the river. One of the most precise anglers I've ever encountered, Burkland fishes many spots that give him less than a twenty-minute window. Initially, it was a bit disconcerting. Our first drop yielded fish in a matter of minutes, but as we pulled in doubles with the fly rods, Burkland was already looking downriver. "All right, time to move on," he said, just a few casts later.

"You don't understand," I told him. "I've been struggling to catch fish all week. This is heaven right now."

"Don't worry, we're going to a better spot."

And he was right. The next drop was nicknamed the Fishbowl. When Burkland killed the engine, he cocked an ear toward the bank, then smiled. Though we couldn't see more than twenty feet, the slapping sound we heard, mixed in with a few cries from a tern, meant stripers. And soon we were into them again. The fish were feeding on young-of-the-year silversides that were leaving the river. We ghosted onto patches of water that were white with the froth of feeding stripers.

And so went the day, leaving fish to find them biting better elsewhere. In between stripers, Burkland talked about

September 11. "I was in some pretty bad stuff in Vietnam, but I can't imagine how bad it must have been for those people to have jumped out of the building," he said.

As we moved down the river, the fog lifted and the scenery went from bucolic to industrial. We passed a salt plant on the riverbank, pipes and conveyors bisecting mounds of white. Near the navy boatyard, a twenty-four-foot runabout motored up on us. A man wearing a Coast Guard–issue life jacket screamed through a megaphone for us to slow down. With three nuclear subs at the dock and terrorism fears running high, the government did not want boats zipping around the waterway. As a precaution, the subs were submerged. "There goes one of the best drops on the river," said Burkland. Apparently the shelf near the navy dock was prime striper habitat. No matter; within minutes we were into fish again.

By noon the day had warmed to the mid-sixties. Burkland stripped down to his shorts, taking advantage of some of the last warmth of the season. It was a bittersweet time for him. "The fishing this time of year is superb, but you know that the stripers are leaving any minute," he said. "One day last fall, the water was crystal-clear. I stood on the bow of the boat and looked down, and below me hundreds and hundreds of stripers swam out of the river, heading for the coast. It was time to go. I just stood there and watched."

5

Massachusetts:
Where the Cows Come Home

During intense nor'easters in Winthrop, Massachusetts, a small suburb of Boston hard on the harbor, waves crashing over the concrete seawall often douse cars parked on Shore Drive. It's not uncommon to find rocks, seaweed, crabs, and even lobsters strewn across hoods. Stripers don't get stranded, but during more pleasant weather you can walk the beach and cast to them with a fly rod. After lunch and a few drinks, I hoped to do that with the renowned flytier and striped-bass specialist Jack Gartside.

I called Gartside when I arrived in Boston during the last week in September. "I'd love to meet with you and discuss the fall run," I told him. "Maybe do some fishing."

"The fall run? I hate the fall run."

"Really?"

"Yes. I call it barrel fishing, or idiot fishing. Any idiot can look for birds. Why would anyone want to fish the fall run?" he said.

"I think it happens to be the best time of the year to be a striper fisherman."

Gartside responded with a prolonged silence and finally said, "Well, I'll meet you at the Winthrop Arms tomorrow at noon, but come prepared with good questions. Bring a tape recorder because it takes too long otherwise. And I hope you've been taking notes because I'm not going to repeat any of this."

It wasn't the warm reception I had hoped for, but I wasn't shocked. Our chat reminded me of something that Lefty Kreh, the grand impresario of American fly-fishing, once said about Gartside: "His paint don't dry."

III   III   III

Gartside vaulted to prominence in 1982 when *Sports Illustrated* ran a feature on him, part of a series about American flytiers. Later, he appeared on the cover of *Fly Fishing in Salt*

Waters magazine straddling an inflatable giraffe as he floated in a lake. One hand held a fly rod while the other kept the giraffe's neck out of his line of sight. His choice of float tube mirrored his tying sensibilities. Gartside throws tradition out the window. And while he, to quote the *SI* piece, "probably fishes more blue-ribbon trout streams than any millionaire alive," Gartside knows more about fishing for striped bass from the shore of Boston Harbor than any man alive. That was why I had called.

Though he warned me that he was often late, Gartside arrived on time. He was wearing ratty corduroy trousers, an aqua-blue button-down shirt with a surplus of pockets, well-worn wing tips, and a Boston Red Sox cap. Thin strands of reddish hair fought off an onslaught of gray.

We took Gartside's regular table, near a window in the back. I could see the green water of the harbor through a gap between two houses. Rain started to fall. Gartside lit an Old Time cigarette, the first of a chain-smoking performance that even my hard-drinking, fast-talking buddies in New York City would have applauded.

I asked him about his introduction to tying. "I was at a sportsmen's show in Boston when I was twelve or so and I saw my idol, Ted Williams. He was tying flies and I bugged him to teach me how. He showed me how to tie a simple woolly worm. I had it for a long time, but it's been eaten by moths by now."

After a stint as a high school English teacher in the 1970s, Gartside drove a cab to support his fly-tying habit. While waiting for fares, he would mount a tying vice on his wheel and create new offerings. His most famous invention is the Gurgler, a simple surface fly made from a piece of foam, a wisp of bucktail and marabou, a bit of Mylar flash, and a small hackle. When stripped it makes the eponymous noise. Fish all over the world fall for the pattern, especially stripers. It resembles nothing as much as anything that makes a wounded commotion on the surface and draws the attention of fish. You can find the Gurgler at any fly shop in the country, but Gartside also sells them and dozens of other handcrafted creations on his Web site.

While on the site you can also purchase a "risqué" fly-fishing print. The prints are essentially French postcards of nude women from the 1920s and '30s. Gartside then imposes either a fish or a fly onto the image. For $20 you can have a five-by-seven of a lovely lady hoisting a striper or caressing one of Gartside's Secret Baby Baitfish flies. The postcards aren't a huge hit in the United States, but the Japanese can't get enough of them. Spurring interest is a column Gartside writes for a Japanese fly-fishing magazine about his postcard creations. It's called *Bizarre Inspirations*.

Not surprisingly, Gartside revels in finding fish where

others would turn up their noses. "I used to go fish over a bubble of raw sewage," he said. "You could find the place in the fog by the smell. It had everything, the whole chain of life. You'd be taking bits of toilet paper and turds off of your line. Everything was brown all around, giant brown bubbles. The stripers loved it."

Our own expedition wasn't over a sewage pipe, but it was close. After lunch I followed Gartside, who drives a red Chevy Corsica, to Deer Island, where we fished just below the sewage-treatment facility. In local parlance it's known as the "poop plant." From the road, the plant looked like something from Planet Remulak—a clutch of giant egg-shaped holding tanks interrupted the horizon, and pipes crisscrossed the entire grounds. I listened for the collective gnawing of microbes at work on the raw sewage.

Once a leaky, overburdened facility that pumped sewer sludge directly into the water, the poop plant—the centerpiece of Boston's recent Save the Harbor efforts—was transformed into a twenty-first-century marvel, at a cost of $3.8 billion. The plant receives waste from forty-three communities and converts it into fairly harmless effluent, which is piped 9.5 miles out to sea in an underground pipe that is twenty-four feet in diameter. It is then released through a series of vents on the seafloor. Though a few environmental groups fought the pipe, no ill effects have been noticed since it opened on September 6, 2000. Workers at the plant like to say they're "number one in the number-two business."

Just hours after the plant began pumping wastewater far out to sea, the waters of the harbor began to clear up. Gartside crossed the raw-sewage locale off his list of top Boston Harbor drops. The mussel beds just offshore of the poop plant, however, still produce fish, but the rain had fallen heavily while we ate lunch and the water was murky.

And though Gartside told me that if he ever returned to this world in the afterlife he'd like to come back as a tern, as I watched him stalk the shore of Boston Harbor with a fly rod, I swore I was looking at an egret. Even in a pair of waders, his six-foot-two, 150-pound frame looked birdlike. As he lifted each leg, slowly sliding his foot over the bottom, I imagined his knees as bulbous black knots. He had a giant wingspan, so that every time he lifted his fly rod to cast, he seemed ready to take flight. Like a winged predator, Gartside was always studying the water, his face cocked to the side, watching the tide play across a piece of structure or observing the way his fly behaved when he stripped the line.

When I set the hook on a nice fish, Gartside asked, "How far out was he?"

"Just outside the mud line," I told him.

"That's where I thought they'd be. Don't be a barnacle. Keep covering ground after you land him."

A few casts later I hooked up again. It was another solid fish. When I released him, Gartside yelled, "What kind of fly are you using?"

"A bucktail streamer with a bit of flash." The fly, which is called Darlene because the tier thought its haphazard concoction of hair looked like his wife when she woke up, was given to me in New Hampshire. It wasn't pretty, but it worked.

"Okay," said Gartside. "I think I'll switch to a fly with a hook." He was joking, but he'd gone without a point of the hook in the past. Gartside says reeling in the fish is not the treat; getting the hit is what matters. "A large part of fishing is fooling fish. Right? We lie to them. We tease them. We hook them. What more pleasure can you get out of life?" Gartside says he's a veritable wizard when it comes to fooling the striped bass of Boston Harbor. "I know if I go to a certain place on a certain tide I will catch as many fish as I want. I know this because of experience," said Gartside, who claims to have fished nearly every foot of the harbor's 183 miles of shoreline. On this day, however, he failed to get a strike, and we soon headed back. He wanted to make it home before the rush-hour traffic.

Standing by the cars, I pointed to a classic fall scene. About 250 yards from shore, birds were diving over blitzing fish. Barrel fishing at its finest. I expected Gartside to launch into a tirade, but he remained calm. "The fall run is a distraction that takes fish away from me on the shore," he said. With a note of revenge in his voice, he pointed to the kayak on my truck's roof and said, "Can two people get in that?"

"No, it's only a one-man boat."

"Have you ever tried to put two people in it?"

"No, but I can assure you that one of us will be swimming if we try it," I said.

"Well, I have a raft in the car. What do you say?"

"Let's do it."

And with that, Gartside opened his back door and worked the raft free. With it came an old dress shoe, the tip of a fly rod, and a water-stained copy of *Billy Budd*. While I was curious about Gartside's change of heart, I decided not to question it. I didn't want to risk setting him off with the prospect of fish now just a good paddle away.

The raft, a rubber dinghy, was a gift from a friend. It was about eight feet long, but it was missing the floorboards, which allow grown men to put their weight on the thin rubber "deck." Gartside pulled two foot pumps from his trunk, and we both began working air into the raft. A woman from a nearby house pulled her curtain aside and peered out at us. Before long the raft took shape. Gartside found two paddles in the trunk and handed me a life jacket. He had a bright orange vest that he threw over his head, and we carried the raft to a set of steps leading down a seawall. We were directly across from the runway at Logan Airport, and every couple of minutes a huge jet roared overhead. "This is nothing," said Gartside. "Before September eleventh the planes landed every fifty-five seconds." I wondered if any passengers looking down would have had a moment of pause when they saw two men launching a raft just half a mile from the runway.

Gartside took a seat in the bow facing me, his long legs drawn up toward his chin. I faced forward, my ass resting partially on the rubber gunwales, and took control of the flimsy aluminum oars. About one hundred yards from shore, I heard the unmistakable hiss of air escaping from the raft. When I looked at Gartside he was studying his pant leg. "I know that's you," I said.

"Who me? What?" he replied, then lowered his head and began hissing. I laughed. "I do that to scare reporters," he said.

"Do you make them row, too?" I asked.

With a plane zooming in at what seemed just above my backcast, a school of stripers surfaced around us, their sloppy feeding sounds silenced by the roar of the 747. I hooked up quickly, and Gartside lost one beside the boat. "I can't resist," he said. "I have to tie on a Gurgler."

Watching him fish was like playing one-on-one with James Naismith. With a snappy strip, Gartside made the fly, well, gurgle. And on the next strip he hooked a striper. "Did you see that?" he shouted. After he landed it, the fish were at least one hundred yards away. As I began to paddle toward them, they moved farther offshore. Waves started to crest our flimsy craft when a pod showed up in the area we'd just left. "I feel like I'm in one of Dante's Inner Circles rowing you around after these fish," I said.

"Dante's Inner Tube," said Gartside, and without missing a beat he pointed to breaking fish about two hundred yards away. "Shall we go after them?"

ııı   ııı   ııı

I left Boston the following morning on State Highway 3 and an hour later crossed the Sagamore Bridge onto Cape Cod. Stripers setting out from Boston Harbor and those joining them from points north face their first proverbial fork in the road when they reach the thin neck of land connecting the mainland to the Cape. There they can either opt for a shortcut through the Cape Cod Canal or cover some open water to reach the northern tip of the Cape, where they'll then swim south along the outer beaches. In the evolutionary time frame, the canal route is a spanking new option.

Though envisioned as early as 1623 by Captain Myles Standish, the canal did not become a functioning reality until 1940, when a previous attempt was set straight, its banks widened, and its depth leveled to thirty-two feet. Soon ships clogged the route, which eliminated the need to sail the 162 miles around the shoal-ridden Cape. Stripers, and even right whales, weren't far behind.

But some fish, like travelers opting for a side road instead of a superhighway, still go the old-fashioned way. Whatever the route, the stripers soon find themselves in the company of the migratory stock's biggest bruisers. Cape Cod is Cow Town.

The surrounding waters have produced more seventy-plus-pound fish than anywhere on the Striper Coast. From a seagull's perspective, the Cape takes the shape of a flexed arm, and sharpies refer to its hot spots in anatomical terms. It's well known that the outer beaches of Cape Cod, from fist to elbow, offer the best shot at huge fish—but not the only shot. I was gawking at giant stripers before making it past Woods Hole, which sits at the bottom of Cape Cod's shoulder.

Many of the most respected pointy-heads in the marine-science field shelve their microscopes in Woods Hole. The Woods Hole Oceanographic Institution, the Marine Biological Laboratory (MBL), and the National Marine Fisheries Service all call this tiny village home. A school of stripers, the sight of which would make most fishermen's knees knock, also calls this area home. I had heard about them from an underwater photographer. "You'll need to call Ed

Enos at the Marine Biological Laboratory," he said. "If you're lucky, he'll let you take a look. But he's pretty protective of them."

From my motel on Water Street, I phoned Enos, who agreed to show me his brood on the condition I remained tight-lipped on their whereabouts. He didn't want to see his finned friends hanging from the scale at the local tackle shop. We met outside the MBL offices, and Enos, carrying a small bucket of fresh squid, hopped in my truck. After a short ride, Enos pointed to a grassy stretch by the water that looked perfect for a picnic.

As soon as Enos's shadow soared over the edge of the seawall, striped bass emerged from the surrounding depths, tails slowly pushing back and forth, pectoral fins feathering the water like miniature handheld fans. I spotted a few fish that were well over forty inches. "This is nothing," said Enos. "A few of them have already left for the fall."

All of the fish, thirty or more of them, waited for Enos to reach into his bucket. When he did, the water's surface rippled with anticipation. Enos tossed an entire six-inch squid thirty feet out. The bass spun like outfielders chasing down a well-hit fly ball and raced to the drop zone. A large cow with a bum jaw opened her mouth just below the surface, and as the water poured in, making a sound like a bucket being plunged beneath the sea, so did the squid. Enos grinned. "She's pretty fast," he said. Then, as if spreading bread for pigeons, he threw out a handful of squid and the water erupted. "Now you can see why I don't let anybody know about these fish."

When the surface settled, I got down on my knees and studied the school. There was a small striper with an open gash on its flank—it probably narrowly escaped the maw of a seal or the twine of a net. The corner of the cow's jaw had been wrenched by a hook. And one striper trailed a string of weeds from its back, hitchhikers on a thin tracking tag anchored near its dorsal fin. Others, it seemed, had made it through life unscathed. The big fish hung back, near the safety of deep water, but the smaller stripers cruised right beneath my feet.

Enos, a former sharpie who has given up fishing for feeding, thinks it's a pretty fitting end for his squid. A superintendent at the MBL, he's responsible for providing the world's top scientists with these eight-legged subjects. Here they're known as the cephalopod *Loligo pealei*. *Loligo* have some of the largest nerve cells found in nature, up to one hundred times the size of those found in humans, which make them the perfect specimens for research. Squid have been essential to the work of the MBL's thirty-seven Nobel laureates.

In the fall, Enos sends his crew out twice a week to trawl for squid. They use a special net that separates the squid from the other bottom inhabitants, such as flounder, crabs, and lobsters. Unlike their West Coast brethren, which school en masse, the squid around Woods Hole tend to hang in groups of twenty-five to thirty. These small numbers make Enos's job a difficult one, "especially when an electrical storm scatters them." A solid day's haul can mean forty squid. "It's a good thing we only have to do this for science," said Enos. When the holding tanks are full, the boat steams to port. The squid go directly to the scientists. Leftovers are stored in giant square tanks on the ground floor of the MBL. Those that don't survive are collected, and some are fed to the stripers.

On our way back to the MBL offices, Enos invited me to take a look around the ground floor. There were circular tanks full of a host of primitive sea creatures—sea slugs, clams, sea urchins, horseshoe crabs, and starfish. After thirty-three years spent collecting specimens from marshes, estuaries, and sounds, I was intrigued about what Enos thought triggered the striper's migration. "It's definitely the sun's rays. These fish show up and leave like clockwork every year. It's October fourth now, and they'll be all but gone by October fifteenth."

III   III   III

Spurred by the sight of these fish, I headed to nearby Nobska Light. Tourists flock to the site to tour the lighthouse and gaze over Vineyard Sound. But I was there to test the waters. I followed a small dirt trail to a jumble of granite boulders protecting the bluffs. I caught sight of a tern dipping into the rip and saw the splash of a feeding fish. As I stepped on a rock that slanted toward the water at about forty-five degrees, my wader boot slipped and both feet went out from under me. The side of my face smacked the granite. When I looked up, the sky filled with little tinfoil bugs, and I tasted blood.

I tried to go on but I was experiencing a thick fog not unlike the one that overcame me when I had my brain rattled as a high school football player. I decided to head back to the car. When I got there, I sat in the driver's seat in my waders and took stock of my condition. My left wrist was severely sprained, if not broken, and my pinkie looked like a purple sausage. I was nauseous and would be for the next few days. Eventually I drove to my motel, one-handed.

III   III   III

After concocting an Extra-Strength-Tylenol-and-prescription-painkiller cocktail, I spent the next day in my motel room, alternating between nausea and fog. I reached a state of clarity the following morning, bought a wrist brace, and set off on my cow hunt. That night I had a trip planned with Tony Stetzko, who owns the Cape Copy Shoppe in Orleans.

He also owns the unofficial Cape record for the largest striper landed from the beach, a seventy-three-pound behemoth. The catch was recognized by the International Game Fish Association as a world record for the twenty-pound line class but was toppled less than a year later.

When I stepped into the shop, I knew I had come to the right spot—three replicas of his big striper held sway over the proceedings, mouths agape, bellies bulging like beer guts. Behind the counter, beautiful homemade wooden lures hung from a cord stretched from wall to wall. A couple of them were a foot long, including a jointed eel plug known as Mr. Wiggly. The three faux stripers stared at it longingly.

When Stetzko emerged from behind the counter, I introduced myself. "I wish you could've been here last night, man. We were into the best fish of the year," he said. Stetzko had taken a couple out to Nauset Beach and found a batch of trophy stripers. Ten of the fish were over thirty pounds, the largest forty-five pounds. All of the fish were taken on eels. "They moved in right at dark; they'll be there tonight," he said. Stetzko felt the fish were part of a wave of big stripers moving down the outer beaches.

On the way to Nauset, Stetzko complained about weekend crowds and spoke about a friend he had to meet that night at nine. "That should give us enough time to bag you a forty-pounder," he said. I looked at my watch. We'd only be fishing for two hours. I could tell Stetzko had other things on his mind.

I wasn't surprised. Stetzko starts fishing hard in May and for much of the season is on the water every night. He augments his struggling printing business by guiding shore fishermen. "Sometimes I practically spend twenty-four hours on my feet," he said. It's no wonder his knees often feel as if sand has worked its way into the joints. To top it off, his marriage had recently run aground, and he was spending a considerable amount of time on a cot in the back of his shop. His mood lifted as we pulled onto Nauset Beach, possibly the most beautiful piece of coastline in New England. We rolled down the windows and sucked in the salt air. To our left, the ocean rolled gently under the southwest wind. To our right, the setting sun cast the sparse clouds in a deep pink.

In *Cape Cod*, Thoreau put his boot to the sand near Nauset and walked 125 miles to Provincetown. So captivated by the sound of the surf, he often resorted to Greek to describe it since, "We have no word in English to express the sound of many waves, dashing at once, whether gently or violently. . . ." Henry Beston came to Nauset in 1926 to spend two weeks in his seaside home, a small cottage behind the dunes, and spent an entire year. Of his decision he wrote in his journal, "The world is sick to its thin blood for lack of elemental things, for the fire before the hands, for water welling up from the earth, for air, for the dear earth itself underfoot. . . . The longer I stayed, the more eager was I to know this coast and to share its mysterious and elemental life." His journal became *The Outermost House*, one of the pinnacles of American nature writing.

Stetzko, who came to the Cape in 1974, took up a rod in favor of a pen, though he dabbles in ocean-themed water-colors. "I don't know where I'd be without surf fishing," he told me. "There's something about figuring out the patterns and movement of stripers which appeals to me. Then you cast out and get that thunk and your line starts whizzing out while you're standing there in the surf. Man, it's a rush, especially with the big fish."

In the fall of 1981 Stetzko cracked the big-fish pattern on Nauset wide open. "The late fall that year had a school of giant bass off the beach for about two and a half weeks. I caught a sixty-one two weeks before, so I knew they were there. The night I caught the record I saw huge swirls as big as a Volkswagen, but I couldn't get them to hit. So I decided to tie on a dropper," he said. A dropper is a small lure or fly that's tied ahead of a larger lure or, in Stetzko's case, an eel. The seventy-three-pounder responded and was so hard to budge, Stetzko assumed he'd foul-hooked it. His surf-casting buddy thought he'd snagged a seal. When the striper finally appeared in the wash, Stetzko instantly knew better. The fish made him a celebrity of sorts on the Cape.

It also made him a marked man. "A lot of people looked up to me after that, and a lot of people followed me wherever I fished. I had to learn to keep a low profile." For many years he rotated beach buggies and went out only well after dark. He did make some money off endorsements, but balked when a reel company asked him to pose with his fish next to an angler hoisting a world-record catfish. "I wasn't going to disrespect that striper by posing next to a catfish," he said.

As we drove along the beach, Stetzko read the water. While the majority of anglers see an endless succession of breaking waves, sharpies notice the subtleties of how the waves crumble and roll. As they fall apart the waves offer clues to the shape and depth of the sandy bottom. Often "holes" form along open beaches. These holes are essentially deep pockets flanked by offshore sandbars. The deeper the hole the more likely that it will hold fish. With the ease of a driver reading street signs, Stetzko eventually pulled to a halt. "This is our spot," he said.

Fifty yards ahead of us a couple in a camper cooked hot dogs on a small grill. "Most of these people know me," said Stetzko, "but I'm bad with names." They greeted him like one would a sports star, offering food and beer. Stetzko eyed the grill. "You mind if I throw some littlenecks on there?" he said, then pulled a potato bag full of clams out of his cooler. The couple responded with delight, and the woman hurried inside to heat up some butter. Minutes later we were wolfing down fresh littlenecks. Apparently, this was a ritual for Stetzko. "This is why I catch fish," he said, his chin dripping clam juice as he sucked one down raw. "Before I fish, I *eat* like a fish."

After the sun set, Stetzko and I readied our tackle. It was getting dark, but turning on my headlamp was forbidden, since the light might spook fish—but more important, it would draw attention to our preparations, signaling others that the master thought the fish had arrived. A rapping noise came from the camper next to us. "You know what that is?" asked Stetzko. "The sound of a good eel fisherman." Baffled, I watched Stetzko grab an eel by the tail with a rag and then unceremoniously whack its head repeatedly on the bucket. "I fish my eels dead or close to it," he said. I grabbed an unlucky eel and made my own awkward beat on the bucket.

At the water we found our place in the lineup of other casters. I landed the first striper of the night. When Stetzko saw it he grimaced. "That's a micro," he said. The fish was thirty inches long, a fine striper by my standards. "As soon as that moon comes up, make sure your eel is in the water," he told me. "That's when the big fish will move into the hole. And slow down your retrieve—it looks a bit fast." On the next cast I turned the handle at what seemed a glacial pace. Stetzko looked over and nodded approval.

A few minutes later, a full-blown harvest moon rose over the ocean. "Worth the price of admission," said a giant of a man fishing next to me. I was watching a satellite that Stetzko pointed out streaking through the night when the big guy reared back on a thirty-pound fish. A woman two people down let out a gasp as she hooked into her monster—another thirty-pounder. The fish were here. Stetzko lost his eel and ambled back to the truck. A few minutes later a fisherman emerged from the darkness, stood surprisingly close to me, and made a cast. I assumed it was Stetzko. With three cranks of his reel, the rod bucked and he was into his cow. Then I heard Stetzko's voice coming from behind me. "Way to go, Dave."

"It's not me," I said. "I thought you were him." Before I could finish the sentence, Stetzko had marched over to the fisherman and let loose with a tirade on beach-fishing etiquette that would have made a mob boss blush. When he was finished, the angler beached his fish, a forty-five-pound cow. "That's my client's fucking fish" was the last thing I overheard Stetzko say concerning the striper.

For the rest of the night Stetzko kept looking at his watch and yelling, "Come on, Dave, time to get your forty-pounder." But it didn't happen. I did manage a striper around twenty pounds, the largest of my trip so far. When Stetzko yelled that I had time for one more cast, I made a wing and offered a prayer. I reeled agonizingly slowly. I imagined my eel snaking in front of a squadron of forty-pound stripers patrolling the surf zone. Nothing. With the eel at my feet, Stetzko looked over and motioned with his thumb to head to the truck. He began reeling in his eel at warp speed. I was watching it slither across the surface foam under the light of the moon when it disappeared into the mouth of a forty-three-pound striper.

III   III   III

The next day my quarry was decidedly easier to catch. I was sitting among a gold mine of littleneck clams on a tidal flat in the back of Barnstable Harbor on the Bay Side of the Cape. With one tug of my trowel, the hard-packed mud gave way. The tines clinked off clam shells. I worked my hands through the mud as if digging for potatoes, pulling littlenecks out by threes and fours until the milk crate in front of me was brimming. In thirty minutes I had unearthed one hundred pounds of clams. Some of them would be served the following day in Boston and New York City's finest restaurants.

The clams belonged to David Ryan, a native Cape Codder. He leased the two-acre mud plot for $50 from the town of Barnstable and had planted his crop two years earlier when the clams were no larger than a baby's fingernail. David's younger brother, Chris, a good friend and one of the best striper fishermen I know, was also out on the flat. I'd met Chris when he ran the fishing department of the Orvis store in New York City. I was editing a piece on saltwater fly fishing and needed to pick up some flies for a photo shoot. Though two or three employees manned the department, Chris had a line of guys in suits waiting specifically for him. It was obvious he knew his stuff. We started talking, and before I left the store we had planned a fishing trip on the outskirts of the city. Since then we've been steadfast friends. When fishing is involved in a friendship, you can skip right past the preliminaries.

Chris grew up on Cape Cod, and striped bass were part of his angling curriculum. During college he began working at Orvis Boston. A rare combination of an unflappable, laid-back demeanor and instinctual fishing skills, Chris was the company's choice to spearhead its saltwater fishing schools. In the winter, during three-day programs in Key Largo, he would proselytize wanna-bes into respectable anglers. Then he would move to Cape Cod for a repeat performance.

Eventually Chris pulled the plug on his fishing career. "I'd been working in tackle shops since I was thirteen. I needed to start making a living," he said. He moved backed to Boston and took a job with a human-resources firm, but fishing is still his greatest love.

After the three of us had dug ten boxes of clams, we

loaded them in the twenty-one-foot flat-bottomed Carolina Skiff and motored to an area marked by white PVC stakes. Dave hopped overboard and handed a giant plastic mesh bag over the gunwales. Inside were at least 125 premium oysters, each one the size of a grown man's palm. Dave pulled out a pocketknife, slit the bag, and removed three oysters. With a deft twist of the blade, he unhinged the shells and handed one to Chris and me. All three of us raised our oysters, as if making a toast, and tilted our heads back. The meat slid from the shell, bringing the salt and chill of the ocean together in a slippery moment of taste. Dave nodded his head in approval and began heaving more bags onboard. "We'll need twenty, so keep count." Afterward, we chugged home, sea farmers with a good haul. For three hours of work, Chris and I were content with our wage—the use of the boat for the afternoon. Like kids, we tore off our waders, grabbed our fly rods, and went zooming out to the mouth of Barnstable Harbor. We were met by a fifteen-knot northwest wind, but Chris found some fish on the edge of a flat, and we spent the afternoon casting flies to school bass.

Late that evening, after mooring the boat, Chris and I drove to Chatham Light to fish the inlet. Fourteen years ago, no such inlet existed. A January gale in 1987 clawed its way through Nauset Beach and breached Pleasant Bay, releasing a deluge of water. Over the next month, four more storms widened the cut. The inlet's mouth is now two miles across. We found the fish less cooperative and gave up when the northwest wind threatened to toss every cast back in our faces. In Chatham, we ducked into a bar called the Squire for a cold beer, even though our hands and ears had long since gone numb. The weather forecast scrolling along the bottom of the TV predicted thirty-knot winds "capable of blowing over lawn furniture." The following day I didn't see any flying chaise longues, but the weather was unfishable. When the blow finally let up a day later, I drove back to Woods Hole to catch the ferry to Martha's Vineyard. I left Cow Town empty handed.

6

Massachusetts:
Islands in the Sound

When I landed on Martha's Vineyard, even the disc jockey on the local rock-and-roll station was gabbing about the Striped Bass and Bluefish Derby. "Hey, anglers," she said, "it's the last week of the derby, and the weather is beautiful. Get out there—I hear the fish are jumping." She was right on two counts; on the third, unfortunately, she was dead wrong. The fish, especially the stripers, weren't jumping. They weren't even really biting. Though resident stripers provided typical summer action, the fall push of fish had yet to show. On the island, talk of their arrival monopolized conversations, whether in line at the bank or over the family dinner table. Old-timers believed the majority of the stripers had begun migrating through the Cape Cod Canal, bypassing the storied island in favor of a shorter route south. More optimistic anglers blamed the stripers' tardiness on the warm fall. Some cited a newspaper report in the *Martha's Vineyard Times* that posited a warm eddy had spun off from the Gulf Stream and pushed colder water near shore, forming a natural deterrent for visiting fish. One thing was certain: With only three days left in the monthlong derby, there was a palpable tension on the island. Every truck bristled with rods. Men in waders walked out of grocery stores loaded with provisions. If the stripers paid a visit, no one wanted to miss them. Derby Fever had spread to epidemic proportions.

Derby Fever, say Vineyarders, can infect anyone from a beautician to a construction worker. Its telltale symptoms include puffy, bloodshot eyes; bait-stained and scale-caked clothes; matted, greasy hair; a thousand-yard stare; and a raging case of insomnia. Locals told me I'd be fine until I bought my derby pin.

III   III   III

The derby is the mother of all striper tournaments. It may also be the most hotly contested fishing tournament on the East Coast. It began a year after World War II ended, as a way to extend the tourist season; the contest would run from roughly September 15 to October 15. That year, bluefish were scarce, as they had been for some time, and the tournament was called the Striped Bass Derby. Anglers could fish from boats or the shore, and there would be a weigh-in every day. Leaders were posted on a giant bulletin board. And while hundreds of fishermen did make the trip across the sound, it was the Vineyard residents who were most inspired by the derby. After years of tackle-shop bragging and bravado, there was a stage for the island's best fishermen to flex their muscles. Rowdy anglers who displayed their big catches on the street and quiet types who never mentioned a word about their exploits were locked in competition. It took little time before cutthroat tactics were considered the norm.

Tailing veteran anglers was considered as dirty as cursing a man's mother, but it happened frequently. Elaborate car chases ensued, and many fishermen resorted to driving the night roads without lights. To lose those in pursuit, drivers slipped off the road into a wooded area as the hangers-on sped past. On the beach, big fish were hidden beneath the sand until weigh-in. Sometimes a decent fish caught on the north side of the island would later be displayed on the south side as a decoy. Even drag marks, the trails left in the sand by the tails of monster bass as they're lugged to beach buggies, were swept away with brooms by diligent anglers.

Many fished from dusk until dawn and ran scouting missions during the day for an entire month. Work on the island slowed to a crawl. Even schoolteachers were nabbed dozing at their desks. Coffeepots ran dry. But after a summer of dealing with nagging tourists, nobody seemed to care. The islanders embraced their native fishermen and cheered them on. Nevertheless, it was a New Yorker, Gordon Pittman, who won the first derby by landing a forty-seven-pound striper. He pocketed $1,000. Second place went to Daniel Huntley of Buzzards Bay. His prize: a chunk of land in Gay Head, a ritzy community on the western end of the island. A tradition of hard-core angling had begun.

In the following years, bluefish, false albacore, and bonito were added to the list of eligible species. In 1985 the striped bass was removed from the lineup due to its declining population. It made a much-welcomed return in 1993.

Nowadays, more than two thousand anglers participate in the derby. It's one of the few island institutions to resist the sweeping changes induced by the Vineyard's reputation as a celebrity hangout. When Bill Clinton arrived for the wedding of Ted Danson and Mary Steenburgen during the 1995 derby, hard-core locals swore that they would march right though the Secret Service phalanx if the suits stood in the way of a blitz. The first shot fired over the bow, however, had come during the 2000 derby when Mark Plante, the eventual grand winner and member of the Derby Committee, was forced from his year-round rental due to skyrocketing housing prices. Indeed, the week I arrived, the front page of the *Vineyard Gazette* led off with a story on the record sale of a single residence on the island for $22 million.

While building lots are no longer awarded to derby winners, prizes aren't shabby. The eight overall winners (made up of the anglers landing either the largest striped bass, bluefish, false albacore, or bonito in the shore and boat divisions) receive rods, reels, paintings, and assorted tackle. Through a random drawing, one of them also wins a nineteen-foot boat and is crowned the grand winner. But the ferocious competition doesn't stem from the loot. Win the derby and you're a celebrity. People whisper about your impressive talents when you walk past, and little kids can't help but stare. Catch the fish under extreme circumstances, like, say, with ten minutes left in the derby, and your story will be told long after you're dead. It's as close to angling immortality as one can achieve.

Getting your pin is the first step. You receive it when you register. I went to purchase my pin at Coop's Bait and Tackle. The owner, Cooper Gilkes, is one of the most respected anglers on the island. The fifty-eight-year-old fished his first derby when he was eight and won the grand prize in 1987. An inside tip from Coop is like a direct link to Poseidon.

Coop's shop is attached to his home just outside of Edgartown. I found it by following a truck with a bumper sticker that read, "Kiss my bass." When I arrived, Coop, of course, was out fishing, but his daughter, Tina, was running the store. She held a corkboard filled with yellow pins in front of me. I chose pin number 4215 for no reason except that it was alone on the corner of the board.

I entered the fly-rod division, because the leading striper weighed only fourteen pounds—a beatable weight—and I had arranged a fly-fishing trip that night with one of the island's young guns. The all-tackle division was led by a thirty-nine-pound bass.

"You were fast," said Tina. "Some guys stand in front of the board for hours. When the little kids come in, Dad tells them to close their eyes and run their hands over the board. He says, 'Let the power of the pin speak to you.' It's pretty cute."

Even though I had two boxes full of flies, I bought $35 worth before I left the store. When three grizzled men stumbled into the shop, Tina looked at them and said, "Dad's not here, guys. He's out fishing. You have any luck?"

The leader of the group, a big man with a thick beard and leathery face, replied, "Not a bit." The guys took a look around and headed out the door. "Did you see the scales on that guy's hands?" she asked.

"I missed them."

"Bluefish. The action can't be that slow."

||| ||| |||

My companion for my first night of the derby was David Skok, a professional flytier and Chris Ryan's former Cape Cod roommate. Often, Chris would wake up in the middle of the night to find Skok in a miasma of cigarette smoke, tying flies beneath a bare sixty-watt bulb that dangled from the ceiling. "I don't know when he slept," Chris had said. Skok seemed like a perfect companion for my first night of the derby, especially since, as he himself so bluntly told me when I met him, "I can't seem to pull the lucky horseshoe out of my ass."

The week before I arrived, the twenty-seven-year-old landed a twelve-pound false albacore that was leading the contest. With three days left, he seemed a lock to take the albie title for the second consecutive year, especially since the false albacore, like the striped bass, were noticeably absent from Vineyard waters. Skok's fish came after a string of twenty-four twelve-hour days on the Lobsterville Jetty and various beaches. After sunset he went striped-bass hunting. When not fishing or fretting about remaining in the lead, he crashed at the cabin of a young guide named Jaime Boyle.

Skok's string of good fortune extended beyond the jetty. Within the past month, the country's largest mail-order fly catalog had added a Skok creation, the Mushmouth, a fly he designed for albies and stripers, to their pages. And former president Bush, who was given a selection of Skok flies when he made a visit to Boston, sent Skok a personal thank-you note. "Dude, my parents were stoked by that," he told me. "I don't think life could get much better."

"Well, a big striper wouldn't be a bad addition to the list," I said.

"Funny you should say that. Tonight's a special night," he said. "It's the twenty-year anniversary of the biggest-ever fly-rod bass taken during the derby: forty-two pounds, thirteen ounces. Kib Bramhall took the fish on the last night of the competition in 1981."

"That's a tough fish on the fly rod," I said.

"What makes it one of the toughest fish ever, other than the size, is that he went out at seven o'clock and fished till two in the morning, and there were only two other schoolies taken by him and a bunch of other guys. Everyone ends up leaving, and he stays. He was too tired to drive home, so he fell asleep in his truck for a few hours. Now, that on its own is not very tough, but what's tough to me is pulling yourself out of your bag at four A.M. when it's thirty degrees out. He gets out and fishes for an hour and a half without a bite. Then all of a sudden, *wham.* Forty-two-pounder."

We met at Boyle's cottage in Vineyard Haven at 7:45 P.M. It was easy to find. A twenty-two-foot flats boat rested on a trailer outside, and the door was wide open. I recognized the decor as soon as I walked in: bachelor fly guide. The large-screen TV was on mute, the computer monitor displayed a marine forecast for Vineyard Sound, a fly-tying bench was covered in spent feathers, flies and plugs hung from every lampshade, and derby plaques and fish prints lined the walls. Boyle had just come in from his thirtieth consecutive trip. He lounged on the futon in his scrub pants and fleece pullover, his hair sun-bleached blond. A pot of pasta was boiling on the stove. We talked strategy for a few minutes, and Boyle gave me some extra-large black flies for the night's adventures. With trip number thirty-one scheduled for 6:00 A.M., Boyle made a wise decision to remain moored to the futon for the night.

Skok and I stashed our rods in the back of his Jeep and headed to the Cumberland Farms for provisions. "This is a fishmobile," said Skok. "The first rule is, there are no rules." My seat was littered with cassette tapes, so I gathered them and moved them to the floor. "Not down there," said Skok. "That'll ruin them. Just sit on them. You won't even notice." I'd broken the only rule in a vehicle ruled by an anarchist.

We entered the store in our waders. Nobody looked twice. Skok grabbed a large coffee, a jug of Gatorade, a bag of Sour Gummy Worms, and a pack of sour-apple gum. I followed suit, and we left like kids returning from a trick-or-treat outing.

Our first stop was West Chop, where a couple of jetties

pushed out from shore into the darkness. It was a beautiful night, and the Vineyard Sound was quiet except for the passing of the ferry. Skok quickly landed a schoolie, and I did the same, then our luck faltered. At Lambert's Cove we had more of the same: nothing. It was close to 1:00 A.M. when Skok let on that he had a gut feeling.

"I don't take lucky fishermen with gut feelings lightly," I said. "Lead the way."

Skok had heard that the herring were starting to drop out of Menemsha Pond and thought there was a good chance some big bass had moved in to feed on them. Somewhere on the twenty-minute ride from Lambert's to Menemsha, fatigue overtook Derby Fever and my head started bobbing. With each awakening snap of my neck, I could hear snippets of stories Skok was telling about derbies past.

As we pulled up, Skok told me about a Vineyard local who once had a dream that a large bluefish was swimming in a pond. The dream wouldn't have been all that strange if the fish were a bass, but bluefish don't like the slow-moving waters of the ponds, and if they do enter they don't remain for long. She woke in the middle of the night and told her husband about the dream. He laughed. "He told her the derby was really getting to her," said Skok. The next night, unable to sleep, she drove to the pond and landed the biggest bluefish of the derby. "Isn't that some crazy shit?" said Skok.

Our first casts didn't produce any bass, or blues for that matter. But as we continued to fan out on the shoreline, I heard a distant splash, like someone dropping a brick in the water. I let it pass without much thought. Then I heard it again, and so did Skok. "Dude, did you hear that? Those could be the monster stripedos we're looking for." We banged knuckles like two NBA players who'd executed the perfect alley-oop and jogged down the bank in our waders. One hundred yards down the shoreline we saw the stripers ripping open the water's surface—an area about twenty yards in diameter looked as if the heavens had opened up and let loose with a cargo of softballs. We waded in and started casting. At times, the fish would rise and crash just a few feet from us, throwing massive amounts of water. I thought of the big cow in Ed Enos's brood. These fish had to be twenty pounds. They may have come into the pond on the trail of herring, but now they were ambushing thousands of silversides. Each surface explosion began with the sound of rain pattering, the silversides leaping into the air for safety, then falling back. Pitter, patter, *sploosh*, the sound echoing off the wall of trees on the bank. We worked the water hard, changing flies and retrieve speeds. "Just too much bait," said Skok. "Must be a million silversides out there. What are the chances a striper will see our flies?" Undeterred, we fished long after the last of the surface activity. Derby fame had been within casting range, but I couldn't hook it.

The next morning, as I stockpiled muffins into a napkin at my motel's continental breakfast, I heard the women in front of me mention a forty-four-pound striped bass. "Somebody catch a forty-four?" I asked.

"Sure did. And the guy's brother landed a twenty-one-pounder on a fly."

"I wonder where they were?" I asked, not expecting an answer.

"They were at the Tisbury Pond cut. Cooper Gilkes sent them there."

III   III   III

If Dave Skok and Jaime Boyle are the Vineyard's new guard, then fifty-three-year-old Janet Messineo is its respected veteran. She's also arguably the most well-known female angler on the Striper Coast. Sure, there are plenty of others, but few with her thirty-two-year history and hard-charging reputation. I had left a message on her answering machine in hopes that she would return my call. She did and invited me to tag along for the night.

"When I first started fishing, the guys used to call me a five-pound shit in a ten-pound bag," said Messineo, as she pulled on her waders. The guys had a point. At five foot two and no more than one hundred and ten pounds, she seemed to be swallowed by the waders in one gulp. When she snapped her suspenders and smiled at me, I was struck by the softness of her brown eyes and had the urge to walk up and hug her. I resisted.

"You ever try those women's waders?" I asked.

"Yes, but they make women's waders with hips so we can look pretty. The problem is, they aren't easy to pull down if you have to answer the call of nature," she said. This can be a major inconvenience with a derby schedule like Messineo's. One of the hardest-working fishermen ("Call me a fisherman," she said. "I think 'fisherwoman' sounds funny") on the island, Messineo had scaled back her hours this derby. It was hard to tell. Each day for the past month she had woken up at 4:00 A.M. to fish for stripers until daylight. Then she pursued bonito, blues, and albacore until 4:00 P.M., took a two-hour nap, and went after bass again until midnight. During her "hard-core" period just a few derbies back, she was pulled over for swerving. The cop thought she'd been drinking; she hadn't. She'd been awake for nearly forty-eight hours. "My husband doesn't really see me for a month. But he knows I can't help it. He's really understanding."

"Does he fish much?" I asked her.

"Not a bit," she replied.

One Christmas, Messineo's husband, Tristan, bought her a Van Staal surf reel. She had wanted one for years but was afraid to make the $500 purchase. Tristam wrapped it in tissue paper and put it inside a pink Victoria's Secret box. "I looked at the box and thought, Uh-oh, what have we got here? When I saw the reel, I started crying," she said.

Her dedication has endeared her to everyone on the island. She's won countless derby prizes and caught the second-largest striper of the 1984 derby—forty-five pounds. When I told Skok that I planned to fish with Messineo, he said simply, "Make sure you have plenty of rest."

We were to meet at 6:00 P.M., but around 4:45 my phone rang. "Dave, hi, it's Janet. I was thinking, if you don't have anything to do, why don't we meet a little earlier than six."

"Sure," I said. "When should I leave?"

"Oh, how about as soon as you can?" After nearly a month of nonstop fishing, Messineo was too anxious to sleep.

When I arrived, Messineo took me for a quick tour of her studio. A full-time taxidermist, she works in a small wooden cottage in her backyard. In 1987 Messineo attended the Pennsylvania Institute of Taxidermy for a nine-week course in skin mounting. There are few marine taxidermists who still work with skin mounts. It's a difficult process that begins with the skinning of the fish. That same skin will eventually be applied to a foam manikin. "Taxidermy," she told me, "is from the Greek words *taxis* and *derma*. Taxis means move and derma means skin."

Messineo's attention to detail and love of fish made her a natural. She always has a freezerful of work; some of it arrives by plane from the mainland and Nantucket. In her studio, it was easy to understand why. Stripers, sea robins, bluefish, and bonito hung from the walls. The fish looked as if they'd been pulled from the water seconds before. There were even earrings made from tiny minnows. In the work area were fish in various stages of dress and even a giant lobster that someone had dropped off for mounting. It weighed nineteen pounds. I picked up an empty claw that was as big as a bread loaf. "The best part of that is, I got to eat that lobster," said Messineo.

"I really love to work with stripers," she told me. "Most people don't realize it, but each fish is different. Some have broad shoulders, some have huge heads, and some are, well, kind of ugly. But I would never tell that to a customer. They love their fish."

We locked the studio and hopped in Messineo's truck. So far, she had landed the largest bluefish caught from shore by a woman. But she had her sights set higher. This was her twenty-fifth derby, and she wanted to win it all. There had been reports of decent fish still lingering around the Tisbury Pond cut, but Messineo thought the area would be too crowded. She hates crowds. It's not unusual for her to leave the fish biting when too many anglers show up on the beach. She also likes to keep a low profile. "Often when I'm fishing I put on my big coat and pull my hood up. I like the guys to think my name is Bruce and I weigh two hundred pounds," she said.

At the beach, Messineo selected a rod from her roof rack

and pulled a five-gallon bucket from the back. It was loaded with fresh squid that she'd spent part of the previous night catching in Edgartown Harbor. Her night's strategy was blood simple: cast out the squid and wait. "That's how I learned. It's what I like," said Messineo, who has no patience for the new breed of fly fishermen who often look down on other methods of fishing, especially with bait. She pulled a squid out of the bucket, caressing it with her hands as if it were a hair ribbon. "Oh, these are real nice," she said.

With the wind howling at twenty knots directly in my face, it seemed fruitless to fly-fish, so I grabbed my surf rod and tied on a black needlefish. We were at Metcalf's Hole, where more than a few derby-winning bass have been duped. Messineo, however, hadn't even weighed in a striper this year. It only fueled her determination. By midnight, I had yet to have a strike, and neither had Messineo. I wandered around a point in the beach and had a seat in the sand. Messineo found me sleeping a half hour later. "Let's get some rest and meet at four-fifteen. I have another spot in mind," she said. On the way to the truck, our scuffling boots caused the phosphorus in the sand to glow. "Sometimes when I'm out here alone and the phosphorus is glowing, I like to sing, 'I have diamonds on the soles of my shoes,'" she said.

A few hours later, Messineo took me to her lucky spot but only after threatening to blindfold me on the way there. I swore I would keep its location secret. I landed a few schoolies on the fly rod, and Messineo caught two bluefish on live eels. After we left, she dropped me off at my truck and drove off to the Memorial Dock in Edgartown where she planned to spend the day pursuing bonito. I decided on a brief nap.

III   III   III

My last shot for derby fame rested squarely on Jaime Boyle. At thirty-two, he'd already been named an Orvis Guide of the Year and held two derby records. But heavy fog and fouled spark plugs in his boat's outboard engine combined to squash our chances for fish. As we sat in Menemsha Pond changing plugs, the radio crackled with news of a capsized boat at Squibnocket. An experienced derby contestant had been striper fishing near shore when his engine conked out. Not long afterward he was caught by a rogue wave that sent his craft crashing to shore. He swam away unharmed, but the news, in concert with the dense fog, seemed to spook most of the fishing fleet. Many boats returned to the dock. Boyle and I joined Skok and the rest of the island's grounded fly fishermen on the Lobsterville Jetty. For most of the time Skok held court, pontificating on fly-tying and women for anyone who would listen. To his relief, there were no false albacore taken. The next day his lucky horseshoe remained firmly in place. He won the false-albacore title for the second consecutive year, and then, by luck of the draw, went on to win the nineteen-foot boat.

III   III   III

The ferry back to Woods Hole was deserted. Most visiting
anglers had left immediately after the contest ended. The
2001 derby would be remembered for the stripers that
never showed and the incessant wind, which was still blowing when I departed. Waves had stacked up in Vineyard
Sound, and as the blunt bow of the 201-foot *Islander* dug in,
spray shot skyward, catching the sun's rays and transform-
ing them to rainbows. As the big ship pitched and rolled, I
climbed into the back of the truck and went rummaging.
When I finally lifted my head I was woozy, but I had found
what I was looking for—a manila folder with CUTTYHUNK
stenciled on it. On the second deck, I bought a beer, found
a nice window spot, and spread the contents of the folder
on the table. In front of me was every bit of information I
had collected on Cuttyhunk—an island that has been drawing striped-bass fanatics since the dawn of sportfishing. I
had my own pilgrimage planned the following day.

Cuttyhunk, a five-hundred-plus-acre hunk of land covered with scrub forests and grasses, sits at the southwest-
ernmost tip of the sixteen Elizabeth Islands. The Elizabeths
rise like a picket fence between Martha's Vineyard and the
coast of southern Massachusetts. Bartholomew Gosnold
discovered Cuttyhunk in 1602. He built a small fort, then
returned to England. The island was considered nothing

more than a sheep pasture until the late 1700s, when it gained a reputation for producing the world's most skilled pilots, including the noted black mariner Paul Cuffee. No doubt the seamen honed their skills navigating the treach erous reefs surrounding the island. In 1864 these same reefs attracted a sailboat chartered by seven New York sportsmen. The men, all millionaires, had left their striped-bass club on Sakonnet Point, Rhode Island, over a flap in the rules and regulations, and were looking for a suitable place to establish a new club. They knew immediately that they had found it. Not only were Cuttyhunk's deep-water reefs loaded with striped bass, but the boulder fields near shore brought the fish within casting range.

The group purchased the majority of the island and began work on the clubhouse of the Cuttyhunk Fishing Association. There would be a total of fifty members; each of them would pay a $400 initiation fee, plus $100 in yearly dues. A portion of the dues would go toward bird food, which would be fed to the club's carrier pigeons, the only source of communication these men had with their New York offices. The rest would go to the upkeep of the club's steamship, wells, icehouses, and gardens.

In no time, many of the most powerful men in America had signed on as members of the new club, including rail-road tycoon Jay Gould and Henry P. McGown, who ran the Hudson River steamship line. John Lynes landed the first striper on June 18, 1864. His catch was no fluke. He would go on to gain admiration from members for his forty-nine-pound cow, not to mention the thirty-three stripers he beached in one day.

The men fished from sixteen bass stands—walkways anchored to rocks by iron pipes and cables that radiated from the island's shores. Young boys were positioned beneath the stands and ladled a mixture of bunker and crabs into the water. The stately fishermen called the boys "chums" and what they did, "chumming." The boys also baited the hook, normally with the meat from a lobster tail, and gaffed the stripers.

After dinner, the day's catch was weighed on the front lawn of the club while the steward played "See the Conquering Hero Comes" on his trombone. The angler with the largest fish won the "High Hook" award, a diamond-studded fishhook pin, until someone surpassed him. Big cows were called soakers.

It was an idyllic life. Dinner was an elaborate affair consisting of many courses. A menu from one summer night in 1908 includes native spring lamb, lobster salad, baked ham, mince pie, and apple meringue pie. And though the island was dry, the men had well-stocked private liquor cabinets loaded with London Gin, Holland Gin, 1836 brandy, and an assortment of rums and ales.

Each man was allowed one guest per week (no women

were allowed), and quite a few notables made the voyage. Presidents Grover Cleveland, William Howard Taft, and Teddy Roosevelt spent time at the club. Most likely they fished from what was known as the "club stand," the closest stand to the clubhouse and the one with the most consistent action. Standing on it, one's view was obscured only by the magnificent cliffs of Gay Head on Martha's Vineyard.

While the club members depended on a steamship to reach the island, I planned to hop the *Alert II*, a sixty-five-foot ferry, out of New Bedford. The day I left was beautiful and clear with temperatures in the sixties. The sky was a promising blue, the type of sky that makes you feel like you can handle just about anything that comes your way. The wind had sloughed to a slight breeze after howling for five days. I parked the truck in a garage near the ferry and explored New Bedford, a town I had thought about since reading *Moby Dick* in high school. During Melville's time, New Bedford was the whaling capital of the world, and leaving a landlubber's life behind was as simple as hauling yourself onto the deck of a ship. Just fifteen years ago, the town was the country's biggest commercial port, but declining fish stocks and subsequent regulations have quelled growth. The town still supports a commercial industry, but its economy has sunk quicker than a ballast stone. It is now a gritty area with a rampant drug problem.

At the wharf, the *Alert II* was dwarfed by commercial fishing trawlers—*Janice Julie, Jupiter, Dinah, Theresa R., Christine & Julie*—docked three deep. The foul weather had forced them inshore, and now the crews were preparing to head back to sea. Women hoisting children on their arms stood beside run-down cars and waved good-bye to their husbands. Most of the guys had greasy ponytails slinking out of baseball caps and wore jeans, hooded sweatshirts, and rubber boots, de rigueur for commercial fishermen. Walking around with a spiral notebook, I was eyed with suspicion.

On the boats, blocks of ice were swung aboard and bags of provisions were handed fire-bucket style from man to man. A few men hunkered down in the engine rooms, black clouds of burnt oil occasionally puffing above them.

At the ferry, things were decidedly more leisurely. The cargo—boxes of mail and fruit, bikes, luggage, two fishing rods, and an Andersen sliding glass door—was loaded first. The weekend service had been canceled due to heavy weather, so the load was larger than normal. When the ferry's horn sounded, I boarded the ship with eleven other passengers and two dogs. Just before the lines were cast off, a UPS truck raced up to the dock. The driver quickly swung down from his seat and tossed a package to a crew member, and we left.

From New Bedford, the ferry tracked south for fourteen miles, and an hour later we motored into the Cuttyhunk harbor, as pretty a place as you could hope to drop anchor in. There are no cars on the island, but a couple of golf carts were parked on the dock. (Thankfully, there's no golf course, either.) I saw fish as soon as I arrived; every wind vane in sight was a striper. I hoped the waters of this island were as thick with striped bass as the sky was.

My host was Jim Donofrio, the Executive Director of the Recreational Fishing Alliance (RFA). At the dock, he was flanked by his two Labrador retrievers, Ruby and Belle. A former big-game charter skipper, Donofrio had established a reputation as one of the best on the East Coast. For twenty-three years, clients had regularly hauled in trophy catches of striped bass, tuna, and, toward the end of his career, massive marlin. Donofrio liked marlin fishing. "The marlin weren't as predictable as the tuna, and there was less science on them, so you always had to be reading the water, figuring out these fish on your own. I found that to be a great challenge," he said. But it was the wanton destruction of this species by long-liners that was the impetus for him to leave the bridge. "I told my friends, 'I'm done with my career, because if I don't stop fishing now and try to make a difference I'm not going to have a future in this sport anyway.' I wanted to create the National Rifle Association for fishing, because nobody was lobbying in Washington like they needed to."

So in 1995 Donofrio established the RFA, and he's fully consumed. Donofrio never goes anywhere without one of his three cell phones. "I need two different phone plans, because I can't get enough minutes," he told me. And it's true—though Donofrio had come to Cuttyhunk for two weeks of vacation, he was always on the line with the office. "Tell them," he said, during a conversation with an RFA employee in Washington, D.C., "we didn't start this fight, but we're prepared to finish it." The talk is typical of Donofrio's operation. The RFA isn't afraid to take stands, whether it's against the government or a coalition of long-liners. In doing so, Donofrio brushes up against some rough characters. He received numerous death threats over the phone while working on a plan to ban long-lining. It doesn't seem to faze him. "It only upsets me because it's cowardly," he said. "It's the lowest form of revolt."

After tossing my gear in my room, I left Donofrio on the phone and went to find the old clubhouse. Ruby, a youthful black Lab, bounded after me. For the most part, the island's twenty-six full-time residents walk everywhere. It wasn't hard to understand why. A velvety sea breeze picked up the scent of wild roses and honeysuckle. Weathered houses were decked in authentic nautical decor: Wooden lobster traps held potted flowers, brightly colored floats lined walkways, and scuttled wooden boats served as benches. Nothing went to waste on the island. Unlike on the more ritzy islands in the Vineyard Sound, the nautical image wasn't forced.

When I came upon the one-room schoolhouse, six students climbed onto the jungle gym. Beneath them a black Lab waited for their landfall. His name was Blue, and he roamed the island, more a town mascot than a stray. When Ruby ran to sniff Blue, a young boy screamed with utter delight, "Oh gosh, there are two Blues!"

The clubhouse of the Cuttyhunk Fishing Association was just down the hill. Though it had recently been turned into a bed-and-breakfast, the structure was relatively unchanged. The two main rooms looked much as they had when the club was in full swing. The place had closed for the season, but the door was open. The club's flag was pinned to the wall. It bore a rocking chair and three stars, which were rumored to have been copied straight from a Hennessey bottle. A large beach-stone fireplace was anchored by two rocking chairs. Along the walls, the shelves were loaded with books. And while there were plenty of Dickens, Cooper, and Irving, the most worn book was a government-produced tome called *Fishes of Massachusetts*. A framed photo showed Teddy Roosevelt and W. H. Taft standing proudly beside a brace of stripers. Near the dining table a frayed copy of the Rules and Regulations was tacked to the wall. They ranged from procedure for choosing stands (they were drawn by number) to gambling ("all round games of cards for money" were prohibited). Even the liquor cabinets, though empty, still had the names of their previous owners printed inside the doors. Down the hall there was a bit of a sea change. Rooms that once housed presidents and tycoons were decorated in pretty pastels, and beds were topped with fluffy comforters and plush pillows awaiting next season's tourists.

After a dinner of fresh lobsters that Donofrio had purchased directly from one of the local boats, we suited up for a trip to the site of the old club stand. The rocks that held it still show the handiwork of the island men who erected the stands. The tops are flattened out, and symmetrical holes that once held iron pipes now collect water and moss.

Under the silver light of a full moon, I worked out as far as possible. At my back, the clubhouse sat on a bluff, rocking chairs visible on the porch. When I felt a puff of cold air, I shivered and cast a glance at the rocking chairs, studying them to make sure they weren't moving on their own. Surrounded by so much history it wasn't hard to imagine the ghosts of former millionaires and presidents still patrolling the grounds, still on the hunt for stripers. Back in the day, members and guests would wake at 3:00 A.M. to begin fishing. In the dark they wielded fifteen-foot bamboo rods. Their reels were made of wood. When a big striper made a run, drag was applied by pushing a leather thong onto the spool. Those who mastered the subtle technique were known to have an "educated thumb."

Instead of a lobster tail, I tossed out my lucky black

Bomber. After a few casts, I hooked a thirty-inch striper. The club's precise logs show that stripers averaged around nine pounds. Mine may have pushed twelve. It wasn't a soaker, but even Teddy would have been proud to haul it up to the scale while the trombone played in the background.

7

Rhode Island:
A Fish's-Eye View

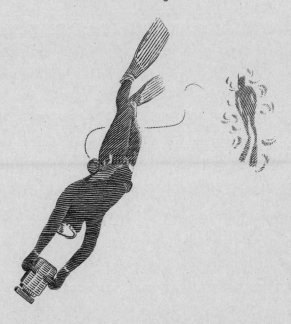

Maybe it was my coastal upbringing, but counting sheep never quite did it for me. When I was a kid I thought about fish when I couldn't fall asleep. Often, I imagined myself a spotted sea trout swimming along the edge of spartina grass, gliding over oyster rakes, zooming by the lug-headed redfish, its nose to the mud, and jetting, mouth agape, toward a school of finger mullet. The loop would run until I fell asleep. An offshoot of my insomnia cure was that I frequently dreamed about fish.

I hoped this preparation would help me when I took to the water with Mike Laptew, who calls himself the "diving fisherman." Laptew left his job as a salesman for a major office-equipment company in 1995 to devote himself to the filming of striped bass. Since immersing himself in his new calling, Laptew has spent more time rubbing fins with stripers than any diver alive. He uses the footage to produce videos that the striper faithful watch with religious zeal. His first production, *Striper Magic*, released in 1995, was a paean to striped bass and their habitat, with plenty of advice for yanking said fish from said habitat. Its tagline read, "The only place you'll see more stripers is in your dreams."

The day I received *Striper Magic* in the mail, I watched it twice. Laptew has produced two other striper videos, *Stripers in Paradise* and *Secrets of the Striper Pros*. One reviewer wrote that Laptew gets "almost as close to striped bass as a sea louse." To provide this close-up footage of stripers, Laptew, who during the early 1980s was one of the country's top spearfishermen, goes in the water without scuba equipment. In doing so, he eliminates the noisy parade of bubbles that spook fish.

There was a specific scene in *Striper Magic* that I had in mind when I called Laptew before my trip. It was a shot taken in the fall off of Rhode Island. In it, Laptew hovers about twenty feet down while in front of him a wall of at least two hundred stripers cruise past, intent on moving south. If you watch it long enough you'll swear you can see determination in their eyes. I wanted to witness a similar scene right next to him. After a few questions about my diving experience (I had little, save for snorkeling in the Caribbean), Laptew agreed to let me follow him underwater.

|||　　|||　　|||

When I arrived in Rhode Island, after hitching a ride from Cuttyhunk to New Bedford on a lobster boat, the wind continued to blow. If wind hinders fishing, it can destroy underwater visibility. The waves tear seaweed from the bottom and escort sand and mud into the water column, creating a murky broth.

I met Laptew at his home in North Kingstown. I found him hunched over his computer in his basement office, studying the live feed from a weather buoy in Buzzards Bay. Laptew has a full head of hair as silver as mercury and a heavy Rhode Island accent. His desk was cluttered with monitors, cameras, underwater housings, and keyboards. There wasn't much light in this subterranean vault, which was decorated with treasures Laptew had picked up from seafloors all over the world: the jaws of a goosefish, an emperor's helmet conch shell, sand coral. There were also faded photos of Laptew holding giant stripers and spearfishing trophies. "Looks like you brought the wind," he said.

That night, while Laptew worked on a side project, a virtual diving tour of Narragansett Bay for the University of Rhode Island, I settled into an easy chair and watched uncut fish porn on his big-screen TV. There were scenes of squid raiding schools of bay anchovies, fluke decimating unsuspecting cunners, eels on nighttime jaunts, and lobster fights. But the striped bass stole the show. I saw them schooling, eating, and hiding behind structure. There was even one segment where Laptew captured a striper by hand so he could remove a fishing hook that was lodged in its mouth.

When Laptew finished up, he told me about life on the bottom. "Oh, it's a glamorous job," he said. "I've swum through cinder-worm sperm and sailfish yak and I've dodged brown pilot-whale asteroids. I may even have the only footage in the world of a jellyfish taking a dump. It looks a lot like a World War II movie, when the PT boats would drop a depth charge." The talk is typical of Laptew's wisecracking sense of humor. He's also a fierce practical joker.

Laptew once descended thirty feet to a rock outcropping where a friend was spearfishing. "I went down right behind the guy and waited for him to settle in. Then I swam just above him and put my finger in his snorkel and wiggled it. He didn't think it was too funny. He started calling me 'the great white asshole,'" said Laptew. Another time, after coming out of the water in Hawaii, he clutched his stomach in front of his friends, doubling over in pain, until a sea cucumber fell out of the back of his bathing suit.

But Laptew, who has lost two friends to free diving, is aware that life on the bottom can be deadly. "To be a good free diver you have to get comfortable with a state of apnea," he said. "The problem is, once you get used to the feeling, it can be intoxicating. The best way I can describe it is like a narcotic. You feel like you could stay down forever, but I'm careful not to push myself like I did in the past." Though he once held his breath at the bottom of a pool for three minutes and forty-two seconds, these days Laptew's dives normally last for a minute and a half.

Down in the wild Laptew melds into his environment. "I tuck into a crack in a rock, let the kelp wash over me. I try not to let fish see my eyes." Most dive sessions last about two hours, and during the season (spring, summer, fall) Laptew's wet suit is always wet. He estimates that he's seen well over fifty thousand stripers. "It's like war, hours of sheer boredom followed by sheer pandemonium." The best scenes capture striper behavior, such as how they relate to structure and current, but the money shot is a blitz. "I've seen striped bass do everything but have sex, and I wouldn't mind watching," he said.

As we spoke, I learned that underwater, the sea is not a silent place. "When a school of stripers gets spooked you hear a *boom*, like a kettle drum," said Laptew. The sound, which scientists often call a thump, is created by the water that's displaced by the sudden tail thrusts and can carry for more than a mile. Laptew told me that he sometimes hears thunderous booms and never sees the culprit, most likely a monster bass that spooked before coming into his field of view. On calm days, it's possible to hear the snapping and clicking claws of fighting lobsters or the gnawing of a blackfish on a crab or a school of weakfish croaking like bullfrogs. For many fish, which can detect minute sound vibrations, the ocean is rent with noise in the fall. Most species, including the striper, can hear a school of bait passing within fifty feet. Through a fish's ears, which are covered by skin, the din of a blitz—with its thumps, slurps, and gnashing of jaws—could be compared to standing at the fifty-yard line during the opening kickoff of the Super Bowl.

We finished talking near midnight and made plans to fish in the morning. Our diving trip would have to wait for calmer seas. I bunked in Laptew's daughter's room; she was at college. I fell asleep among drifts of teenage clutter— prom photos, homemade collages, teddy bears, and a flock of fluffy pillows.

At 4:45 A.M. Laptew woke me up with hushed excitement. He'd been studying the weather reports on the Internet and thought we might have a two-hour diving window when the wind died down between 6:00 and 8:00 A.M., before it shifted to the west. "I don't know what you do to get your constitution up and running, but I suggest you do it," he said. "Coffee. Coke. Whatever works. Get down twenty feet and between the water pressure and your tight wet suit you'll have to answer nature's call if you haven't already. It never fails." And so, we went about our business.

In the car, we headed southwest to Weekapaug. It's one of Laptew's favorite hunting grounds, especially in the fall, when stripers seem to be moving by in an endless procession. To help us relax, a requirement for breath holding, Laptew put in a CD by the Swiss harpist Andreas Vollenweider. Whenever we passed a stretch of water, Laptew launched into intricate detail about the bottom structure of the area. At Weekapaug, the sun was pushing over the ocean and birds were diving on bait just offshore. Two flats boats were working the fish. As predicted, the wind had backed off, but high cirrus clouds told of its imminent return. "Time to put on our Captain America suits," said Laptew.

Before I pulled on my wet suit, Laptew squirted dishwashing soap into it. "Don't worry, I'm not trying to tell you anything. Just makes it a lot easier to slide into." To counter the buoyancy of the thick neoprene, I strapped on a thirty-pound weight belt. I also donned a hood, booties, and gloves. We planned to work the shallow water first, giving me a chance to get used to the conditions, then make our way to some rock piles in twenty-five feet of water.

After a few kicks, I was reminded this wasn't the cerulean water of the Caribbean. This was more like swimming around a farm pond. Bits of green algae—what Laptew calls "snot"—and larger pieces of weed that resembled chewed lettuce were suspended through most of the water column, limiting visibility to four feet. In the shallows, boulders seemed to emerge from nowhere and strands of seaweed rose from the bottom like the arms of a sea creature. When Laptew veered offshore, I gratefully followed. Soon we were kicking up and down swells.

With both hands, Laptew held a digital video camera ensconced in a waterproof housing. Tied to his dive belt was a thirty-foot yellow polypropylene rope with an orange float at the end. A dive flag was planted on the float. For forty-five minutes, I struggled to keep up with the flag while Laptew executed numerous bounce dives, quick trips to scan the bottom for life. Though he was clumsy on land, Laptew moved beneath the surface with the grace of a pinniped. He glided effortlessly to the bottom, bending and twisting as if his bones had turned to cartilage. "Humans can't fly, but when we're neutrally buoyant in the water we can come pretty close," he told me the previous night. When he settled prone on the seafloor, he resembled nothing more than a lumpy rock, save for barnacles and seaweed.

Though I started competitive swimming when I was just

five years old, I was far less graceful in the water and quickly gave up on my own bounce dives. I felt more like a child racing to pluck a quarter off the bottom of a pool. When I did settle on the seafloor, my lungs urged a hasty retreat and I kicked to the surface. I tried to embody Laptew's advice, "Relax, relax, relax." I did so by remaining on top and chasing the dive float.

Even so, my legs had started to burn and I'd developed an unbelievable urge to urinate when the fish call came. Laptew rose from one of his dives, spit out his snorkel, and yelled, "Go down, go down!" and then slipped below the surface.

At twenty feet, the bottom suddenly rose to meet me, a mix of sand and small boulders. I held my nose and blew out to pop my ears. There was less sediment in the water at this depth, but the low visibility gave the impression of swimming through a green-tinted fog. Laptew was on the sand, his camera pointed toward at least twenty-five stripers. They were all similar in size, around twenty-two to twenty-four inches. They moved through the water like shafts of light. They were active and on the hunt. I was so excited I pointed at the fish. The sudden movement spooked them and, like loaded springs, they vanished—*boom*! Then they were back. A wall of stripers.

I watched as they cruised just above the reef and recognized their seemingly haphazard swimming pattern as a shoal—a group of fish behaving independently of one another, not tightly packed and swimming in a single direction. Shoaling fish radiate from a central spot as they explore their surroundings. When I rose to the surface, the fish, sensing danger, formed a school, instantly aligning in one direction and packing in close. This behavior presents a large flank to the predator that hopefully confuses or frightens it. When I returned to the bottom, the fish were shoaling again. Scientists aren't in total agreement on the purpose of a shoal, but most think it makes it easier to find food, since more ground is covered. As with a school, a shoal also provides safety in numbers. The speed of the shift from shoal to school was amazing.

The mood in this shoal was relaxed; at least that's what I inferred, after hearing Laptew's lecture on striper behavior the previous night. The first dorsal fin of a striper behaves like a deer's white tail. When a striper is feeding or senses danger it throws up its dorsal. This school kept their fins down until I made my awkward moves to the surface—*boom*!

When I reached the bottom on my fifth trip down, the stripers had vanished. Laptew, who had been resting on the seafloor when the fish skedaddled, told me they had headed south. We scouted a few more spots but found nobody home and called it a day.

Laptew had about a minute of decent film, but none of it was video quality, due to the poor visibility. In his office, however, he hooked his camera into the computer and we watched the footage. "Most of the time I have no idea what I've captured. When I'm down there I'm just trying to blend in," he said. It was quickly apparent that the film had caught much more than I witnessed. One fish glided up to the camera and pivoted with a quick flip of its tail. "Wow, that's nice," said Laptew as he rewound the film and played it over and over, like a high school football coach analyzing game tape. Play. Rewind. Play. Rewind. When we saw a striper turn on its side to eat something, we repeated the process, same with the lone bluefish that momentarily appeared in the frame. In my cameo, I descended to the bottom like an ungainly monster in a B movie. When we finished viewing, Laptew made me a personal video of our trip, dubbing in music and adding the title "Dave's Most Excellent Diving Adventure."

||| ||| |||

Stoked by our successful trip, and relegated to sightseeing by a thirty-knot west wind, we decided to take a tour of local striped-bass highlights. Laptew began by driving to Jamestown, a small island in the mouth of Narragansett Bay. It has a regal striper history. In the late 1800s Jamestown sported some of the first bass stands in the country, and the names of its top drops—Bass Alley, Bass Rock, Lucky Strike,

Dawsons Stand, Lions Head—quickly became part of Rhode Island vernacular. On October 22, 1936, Arthur Clarke, a local plumber, landed a sixty-five-pound bass from Hull Cove on the island's southeast shore. It was the largest-ever surf-caught striped bass at the time and second overall to Charles Church's world record seventy-three-pounder landed in 1913. Of note also, Clarke felled the giant with a lure he had recently patented, the Eel Bob, a lead collar and hook applied to an eel's headless body. The added weight made for ease of casting and also sent the lure to greater depths. Clarke became Rhode Island's striped-bass guru, and his lure business boomed. He even appeared in an August 1949 *Saturday Evening Post* feature story about surf fishing for striped bass, entitled "No Sport for Sissies." The author called Clarke "the acknowledged seer of striper fishing around the island."

Laptew stopped the truck at the Beavertail Lighthouse, on the southern tip of the island. Gusts of wind rocked the truck as large swells exploded on the shore. When I stepped outside, my hat went zinging off of my head, and I chased it down like a character in a silent movie.

The mount of Clarke's record fish now resided in the lighthouse's museum. "I think it made the rounds between a few tackle shops and then was discovered in somebody's basement in pretty poor shape," said Laptew. "It was refurbished, but whoever did the work was a little overzealous. It has a few bonus stripes." The museum was closed, but Laptew led the way through some bushes and we peered at the striper through a window. I counted at least eleven stripes (four more than customary). They stretched across the fish in a series of large, unrealistic dots and dashes. "Kind of looks like Morse code," said Laptew.

Laptew's own striper history was interwoven with Beavertail. "This is where my one-armed cousin taught me to surfcast," he said. I snuck a glance at him to make sure he wasn't feeding me a line. He wasn't. Beavertail is also one of his favorite spots to dive, because the bottom drops away "like a mini–Grand Canyon" just yards from shore. The intense relief provides a haven for game fish. It also allows ships of all sorts to access Newport. While diving the area one morning, Laptew noticed a "pinging" sound every few seconds. He rose to the surface to see a submarine's conning tower heading for him.

"I live where I live because my house is nineteen minutes from where I worked for twenty-three years and nineteen minutes from Beavertail," he said.

After leaving Beavertail, Laptew pulled into a dirt driveway and parked in front of a ship's wheelhouse resting at the driveway's terminus. "Let's see if Art the third is home," he said. Arthur Clarke III was the grandson of Arthur Clarke Sr. We found him tinkering inside of the wheelhouse, a bear of a man with heavy features. He was wearing blue work pants, a plaid shirt topped by a brown cable-knit sweater, and a blue hat. When he saw Laptew, he said, "Hi, Larry. Nice to see you." Laptew didn't bother correcting him.

The wheelhouse, which Clarke had restored, came from the ferry *Hammonton*, which ran between Jamestown and Newport before the bridge was built in 1962. "For a while, they had the mount of my grandfather's fish in a glass case on the automobile deck," he said as we walked to his garage.

The garage was a trophy room of sorts for his grand-father's catch. On the cinder-block walls, next to a fireman's ax and a sign advertising the Eel Bob, were framed newspaper clippings and photos of the catch. In the pictures, Clárke Sr. looked every bit the "hell-for-leather surf fisher-man" described in the *Saturday Evening Post*. There was also a clipping from 1963 of Art III and his grandfather posing next to a 48¼-pound striper. Art III caught the fish at Brenton Reef in Narragansett Bay aboard his uncle's boat and, as the newspaper reported, "re-established a family tradition."

It was, however, the last catch of note that Art made. He hasn't lifted a rod since 1978. "My business started taking off and I got involved in the wheelhouse project and I guess I just never went out again," he said. But when he unfolded a map of Jamestown fishing drops, the names seemed to jar loose a quiver of good memories. "My grandfather would have laughed at guys going out in the surf these days and catching two or three fish. He told me the year before he died that 'he'd seen the best of times.'"

||| ||| |||

Doubtful Clarke Sr., once "the acknowledged seer of striper fishing," could have foreseen our next stop: a date with Laptew's buddies from StripersOnLine.com (SOL), a Web site catering to bass fanatics. Before we left for our tour, Laptew decided it was a perfect night to stage a fling, what SOL members call a gathering where they meet, eat, drink, swap stories, and often fish. This fling would involve all but the fishing. The chosen spot was the Coddington Brew Pub in Newport, where the house beer contains blueberries.

Laptew had baited his E-mail invite with a powerful one-two combination, promising not only that he would bring new footage but that John Haberek would be there with his latest custom-made striper plugs. We planned to meet at 7:00 P.M. so that anyone wishing to fish the full-moon high tide at 9:30 could fraternize for a bit and head out.

The Internet has freed taciturn sharpies to talk with striper newbies, and SOL is where most of this chatting gets done. Guys who could stand next to a fellow surf fisherman all day long and only offer an occasional head nod now tell their cyber buddies everything from their latest fishing

137

exploits to their niggling complaints about a neighbor who won't stop blowing on his bagpipes.

When we arrived, Laptew spotted a few trucks in the parking lot with fishing rods locked to their roofs. "Looks like the gang's here," he said. Inside, where the yuppie decor seemed more suitable to a gathering of sailing skippers, Laptew told the hostess, "We're with the group that smells like fish." At the table, everyone went by their handles from the site. The group consisted of Crafty Angler, Tattoo, DZ, and Hab. There was also a fellow—Iron Mike— who didn't belong to SOL. According to Laptew, these guys were among Rhode Island's most hard-core striper fishermen. Haberek, who was built like a linebacker and sported a thick handlebar mustache, was the center of attention.

When I sat down, he pulled out a lure box full of varying-size needlefish, long plugs that range in size from the width of a broomstick to that of a pencil. Iron Mike and Tattoo (he has a large striper tattoo on his back) started rubbing their hands together. Most sharpies feel Haberek's needlefish have no equal when it comes to seducing big bass. When stocks run low at tackle shops, anglers will often call Haberek pleading for a personal shipment. "You won't believe the things guys will do for these plugs. I had to stop giving out my number," he told me.

When Haberek opened the box, Tattoo snatched up a

chartreuse needlefish and ran it under his nose like a fine cigar. "Oh, this is real nice," he said. Cries of "plug ho" went up around the table. "Plug ho" is a term of derision for any member of SOL who displays loose morals in the quest for a hard-to-come-by plug.

Many of the guys at the table tinkered with plug making, and they peppered the master with questions. The discussion then turned to favorite lure colors. If stranded on a deserted island (but surrounded by stripers) with one plug, white was the choice for DZ, the wise elder of the group. Most others followed suit, but natural wood and yellow pulled a close second. "Most lures are painted to catch fishermen and not fish," said Haberek.

Everyone at the table was jazzed about the fall run. It was proving to, be the best in recent memory. The bait had stayed close to shore for the past few weeks, and daylong blitzes were common. When I told the guys that the fishermen on the islands in the Vineyard Sound were having a meek year there was little sympathy. Many of them felt the abundance of bait near the Rhode Island shore had attracted the migrating fish. Iron Mike, a spark plug of a fellow who got louder with each beer and seemed to somehow work "fuck" into every sentence, had landed a fifty-one-pounder on a Hab's needlefish just a few days prior. "It can be boom or bust in the fall, and we're fucking booming this year," he said.

I asked the group about the sharing of information on the Internet. Many fishermen I had met on my trip abhorred the idea of posting secret strategies and drops on fishing Web sites. This gang was no different. General information was fine but posting a report with the exact location of where you whacked the fish was a deadly sin. "We all had to pay our dues. Why should some lurker who doesn't know squat have the privilege of being told where to go and what to use? The learning process is part of the sport," said Tattoo. Some members had even devised a code for exchanging information. The following day Tattoo sent me an example. It read:

> The patturn is open, your cleared for twinkies. It was 20 times b4 puke. Pink was the color of the sky in 1977. 7 times I like pink. I like to swim metal. 12 was a good year. Pukers like steel 6 times.

He hadn't written the message, nor was he the intended recipient, but he helped me make some sense of it. Our attempt at a translation read:

> Any number of lures worked, but most of the stripers were small. I caught twenty before the tide changed. I was fishing at dusk at beach area seven. I used spoons. My biggest fish was twelve pounds. I also caught six bluefish.

As the party shifted to the bar, Laptew persuaded the bartender to play my dive tape. I grabbed a stool next to Tattoo, who bears an uncanny resemblance to Ben Affleck, and asked him about the eponymous artwork on his back. "It's an extension of me. Fishing for striped bass was an obsession that became a lifestyle."

Though he was reticent to show me his tattoo in public, he relented when the entire gang started egging him on. When he lifted up the back of his shirt, shouts of encouragement went up from the group and another round of the blueberry sauce was ordered. A turning striper covered most of Tattoo's left shoulder blade; its head, mouth open and gills flared, just touched the crevice made by his backbone. In front of the striper, six peanut bunker swam desperately toward Tattoo's right shoulder. I'd seen plenty of striper tattoos on my journey, but most were off-color and no larger than a silver dollar. This was truly impressive.

"It took three and a half hours," said Tattoo. "The first hour was okay. The second hour I knew something was going on. And the third hour was really intense shit. There was an Elvis clock in the room, and I saw Elvis's legs moving, but time seemed to be standing still. Thankfully it ended just when I was ready to say, 'No more.'"

"Were you worried about the quality?" I asked.

"Not at all. The guy had done work on one of the Allman Brothers, so I knew it would be good."

Midway through the night, DZ bowed out. He'd been

doing rather well lately at one of his honey holes and he didn't want to let the good moon go to waste. The rest of us continued to drink and talk about striped bass.

When Haberek left, he handed me the lure box. "Give these a try and let me know how they work," he said. "Just don't show the other guys." He also invited me to fish the Quonochontaug Breachway—"Quonnie" to regulars—in Charlestown the following night. Some of his friends rented a house near the breachway for three weeks each fall, and word was the stripers were thick.

When the rest of us readied to leave, Laptew grabbed the cassette case of my tape. "Everybody's a comedian," he said and handed me the case. The cover, on which I had written, "Dive w/Mike Laptew," now read, "MUFF Dive w/ Mike Laptew."

lll   lll   lll

When I arrived in Quonochontaug the next afternoon, I noticed some gulls pinned to the sky, but houses blocked my view of the ocean. That the birds weren't winging in a general direction gave me pause—there could be fish under them. Curious, I parked the truck in an empty carport, hoping my out-of-state plates would give me a bit of leeway if the owner arrived. I grabbed a rod and my surf bag and scurried up a beach path in a pair of rubber fishing boots. There in front of me was the first all-out blitz of my trip.

This wasn't a pod of fish attacking a school of bait. No, it looked more like every striper in the sea had decided to lay waste to anything it could eat. Feeding fish and birds stretched the entire horizon. When waves rose up to break, stripers ghosted through their midsections, dark masses of silversides and peanut bunker fleeing in front of them. In the shallows, bass, their dorsal fins slicing the surface like tiny sails, chased scores of peanut bunker. Many of the peanuts were driven onto the sand, where they lay like a handful of shiny coins before gulls hopped about picking them up.

I scrambled atop a small jumble of rocks and sent my Kastmaster into the melee. A fish tugged back before I could tighten up the line. It was a schoolie, close to twenty-two inches. It happened again and again. I switched to a Hab's Needlefish and the stripers walloped it. On one cast I even hooked two bass at the same time, one on each hook.

After landing thirty or more stripers, I went back to the truck, grateful it hadn't been towed. Down the street from the beach, I pulled into the driveway of a one-story home with a white picket fence. There was only one car in the driveway, but rods and waders were everywhere. I peeked into a cooler by the porch. Two huge bass lay under a bag of ice. A man appeared in the doorway wearing long johns and a sweatshirt. "Hi. I'm here to meet Hab," I told him.

"Big fucking deal," he said. "Come on in. I'm Jack. Hab isn't here yet." Inside, more rods lined the walls. "These guys keep the place a mess," he said. "Like a bunch of kids." As soon as I told Jack about the blitz, the two of us left for the beach, but the fish were down. A little while later Haberek showed up carrying a homemade pot of beef stew he had made that morning. Two more fishermen followed him in. As we talked, a guy groggily stepped from a bedroom. He had fished all night. "Ah, sleeping beauty," said Haberek. "Dave, meet Kansas City Andy." Kansas City Andy owned a tackle shop in Kansas, of course, and had come for the fall run. As we spoke, a man named Don sauntered through the front door with a brown paper bag loaded with tackle. Don owned a tackle shop nearby. He had made a run to restock the house. Like Santa Claus, he reached in the paper bag and tossed items to the guys, who oohed and aahed at their gifts. When a spool of fluorocarbon leader, a relatively new material that's extremely effective but expensive, went in the air, three guys made a leap for it. "Hey, you used at least six feet of mine last night."

"Bullshit. That wasn't yours, anyway."

After everyone ate Haberek's stew, we readied our tackle. Though it was 9:00 P.M., we wouldn't begin fishing until the tide started going out, around 1:00 A.M. The lure of choice was a Gibbs black darter fished with a small black teaser. When I put my fluorocarbon leader on the table, Jack pointed to it. "Be careful. I wouldn't leave that sitting out around here."

At 12:30, with the rest of the house involved in a heated poker game, Haberek suggested we head out to the jetty where we would fish. We stepped outside into a clear, cold night—there was a frost warning for the area. As we slipped through backyards to get to the breachway, our breath formed clouds that betrayed our whispers. I asked Haberek about the lure business.

After a work injury in which the chute on a cement truck had come loose and crashed on his back, Haberek had decided to spend some of his free time making a plug. "I had no experience with wood. It would be like you calling me today and saying, 'Hey, Hab. I need my gallbladder out. Can you do it for me?' And I would say, 'Sure, come on over.'"

The first lures he made weren't pretty or functional. Most of them ended up as kindling in his neighbor's fireplace. "I remember I made two casts with one of my early plugs and it came back a mess. It looked like cracked corn. I did everything wrong for a couple of years. It made me sick to my stomach, because I was blowing through money. My wife was saying, 'Come on, you better get this working.'"

Haberek eventually worked out the kinks, and when he

started handing out lures to friends, an amazing thing started to happen. Like wooden sirens, they seemed to entice stripers to strike when nothing else would work. "Guys were actually killing the fish with them," he said. "One charter-boat captain on the Cape called and said, 'These squid poppers are slaying the stripers.'" Word spread, and soon Haberek couldn't keep up with the demand.

As we approached the jetty, we shifted the rods on our shoulders and concentrated on the jumble of rocks at our feet. The breachway, formed by two jetties about fifty yards apart, allows an exchange of water between the ocean and two salt ponds. On full-moon tides, the water rips through the passageway. Fall in on an outgoing tide and you'll be one hundred yards out to sea in no time. Falling *on* the jetty, however, is much more common. A mishmash of jumbled rocks, the place becomes a minefield at night. Not only are there holes, bumps, and crevices, but many of the granite boulders are slick with algae growth. The regulars call them "Teflon-coated." To combat this, anglers tie Korkers, rubber soles laced with spikes, to the bottom of their boots; some guys wear golf shoes. Don told me he once found a man wedged in a hole in the jetty. He was so fat they couldn't budge him. "He was jammed in there," he said. Don called the cops, who eventually pried the man loose. Once out of the hole, it was discovered that he'd broken both of his legs.

Our trip to the tip of the jetty was less eventful. Haberek

schooled me on Rhode Island breachway fishing. You joined a line, and when it was your turn you were allowed one cast into the prime area. When your lure hit the water, you paid out one hundred yards of line as it shot back in the current. With your lure far enough out, you stepped to the left and another guy moved down. He would cast only after you yelled "Clear" to signal that your lines wouldn't tangle.

At the tip of the jetty, the line was seven deep. Nobody said much. A few guys smoked cigarettes, while others simply stared out to sea. When my turn came I carefully stepped down to the last rock, using the butt of my rod as a cane. My knees shook. On the flat of the rock I made a cast. Haberek hopped down with the confidence of a billy goat. "Take your time down here. This is your rock now. Get comfortable and don't worry about the other guys," he said. With my line sufficiently out I moved to my left and Haberek took my place. I could feel the current working against my darter. With my lure almost in I moved farther to the left. Haberek now stood in the spot I vacated. He set the hook with a grunt. A few minutes later he lipped a twenty-pound bass. On my next cast I did the same. For the next hour, almost every angler set the hook on a striper or missed one. My largest fish worked the current like a kite in the wind. As it neared the jetty I inched down to the water's edge to land it, but the fish made one last run, the tip of my rod suddenly springing free of its load. My striper was gone.

When I reeled in my darter the middle hook was missing. The striper had wrenched it loose.

By 3:00 A.M. the action had slowed and Haberek and I headed home, the metal spikes of our Korkers clicking on the jetty. My hands and toes were numb. The following day I would leave for New York. At the house, Haberek crashed on the couch, and I found a spot in the corner, where I hoped to be clear of guys stumbling home from fishing, and tucked into my sleeping bag. I used a pair of wader boots as a pillow. And though I didn't need to imagine myself a fish to fall asleep, I had fish dreams all night.

New York: Gone Skishing

The E-mail from Paul Melnyk arrived on my last day in Rhode Island.

> Montauk has been alive for the past few days! Bass and blues are on the migration, with many 30's falling to the hook in the wee hours of the night. The stripers are fat and dark black on top, with purple pink withers ... new batch of fish. Where are you? Time to go skishing.

The promise of big stripers on the move was encouraging, but the reality of the words "wee hours" and "skishing" gave me a case of the jitters.

I had contacted Melnyk a few weeks earlier because he's the most extreme fisherman on the Striper Coast, most likely the entire East Coast, and I wanted to fish with him. Skishing is the term he coined for his method of fishing—swimming into the sea in a wet suit and casting while he drifts with the currents—because large fish drag him around like a water-skier. Back when I called Melnyk, swimming in the ocean during the dead of night with my surf rod and a batch of live eels seemed like good fun. As I neared Montauk on Route 27, the ocean, rolling and endless, to my right, I began to question my sanity.

III   III   III

There are reasons sharpies call Montauk their Mecca. It sits on the farthest reaches of Long Island, which juts almost perpendicularly from the East Coast, virtually guaranteeing that any southbound fish practically rub its shores as they pass. Just offshore, currents from Long Island Sound collide with eddies spinning off of the Gulf Stream and the semidiurnal cycle of the ocean's tides. This wash cycle is enhanced by dramatic humps that rise from the seafloor, forming what anglers call rips. When a volume of water is squeezed over a hump, it speeds up, as if being forced through a hose, disorienting bait and delighting all manner of game fish, especially stripers. But even if you're not skishing, fishing here isn't for the faint of heart.

The epicenter of surf-casting activity lies just beneath the Montauk Lighthouse, where rats the size of Jack Russells dart among the riprap. On any given weekend in October you'll find fishermen lined up shoulder to shoulder day and night. For the most part, camaraderie is not in their vocabulary. Cross or tangle someone's line and you'll get an earful. When a well-known TV fishing personality was filming a show in Montauk, his production trucks caused a bit of a jam on the beach. Locals let him have it; he swore never to return.

The attitude doesn't get diluted offshore. Each morning, hundreds of charter boats motor out of the inlet, turn east, and head, like a giant convoy, to the waters near Montauk Point. Once there, they troll parachute jigs through the rips with wire line. The wire helps get the lures down deep but offers no sport when fighting the fish. Within minutes, huge twenty- to forty-pound stripers are being hauled to the boat. Not surprisingly, the rips are congested. Smaller boats often get muscled out of the way by larger crafts, and if a private boater makes a mistake—say, getting in the way of a charter guy—the marine radio erupts in profanities. In the worst cases, the larger boat steams into the offending captain's lines and spools his reels.

Even the fly anglers, a group known for their civility, have been known to lose their cool. The number of fly-fishing guides in Montauk in recent years has tripled, and when a pod of breaking fish is spotted, boats converge like children on an ice-cream truck. Most often the roaring engines send the fish down, and those who arrived first sometimes shout a volley of curses at the latecomers. The ruckus reached such a pitch that the normally taciturn fly-fishing captains started a guides' association in an attempt to foster friendships and not feuds. Lately, the guides have been encroaching on the surf gang when the stripers are near shore. In response, surf casters launch their lures right at the boats.

Still, anglers flock to Montauk because the stripers do. In fact, every fish on a fall migration seems to stall for a bit at Montauk to join in a giant feeding party. The din reaches a crescendo in mid-October, when schools of blitzing stripers, bluefish, and albacore pave the ocean's surface with froth. I, like thousands of other fishermen, invited myself to this wingding, and according to Paul Melnyk I was arriving right on time.

III   III   III

I first saw Melnyk at work in the waters off Montauk in 1998. It was a bluebird day—sunny, warm, and calm, just about the worst conditions you could have in the surf. And as surf fishermen are wont to do when things don't look promising, they had gathered in little groups around their beach buggies, chatting like ladies at a tea social. Not a regular, I stood by myself and quickly noticed that their eyes were focused about three hundred yards offshore, where a man clad in a wet suit and white Gilligan-style hat worked a surf rod. He was rising and falling on the waves like an untethered lobster-pot buoy. I heard a chorus of groans as his rod began jumping and quivering. "Son of a bitch," said an angler peering through his binoculars, "he's got another one." Angst quickly spread through the pods of anglers, reaching its apex when Melnyk turned to the crowd, hoisting a twenty-pound striped bass over his head like a prizefighter's belt.

After witnessing Melnyk that afternoon, I went back to work and immediately got him on the phone. I told him I wanted to write a story about skishing for a magazine where I was working. He was more than happy to oblige. It was a quick one-page piece, but it garnered more attention than most anything I'd ever written. Out in Montauk, however, hard-core surf casters weren't too fond of Melnyk's skishing technique. Most anglers thought it was cheating; others thought it was plain stupid. It didn't help that the Coast Guard twice scrambled to "rescue" Melnyk while he was happily catching fish off of the point. And one early copycat did get swept out to sea in a tidal rip. About a quarter mile offshore, the unlucky fellow, thinking fast, dropped his lure to the bottom, where it snagged and prevented him from drifting to the Azores. He was eventually plucked from the ocean by a commercial fishing vessel.

The brouhaha reached a climax during a local fishing tournament in the fall of 1998. The tournament committee met and decided Melnyk held an unfair advantage over other casters who were limited to their waders. Melnyk filed a grievance but lost.

The feud was still flickering between Melnyk and other locals when I met up with him on a crisp, clear Monday. I had arrived the previous afternoon in time to make a quick run to the point with my brother Stephen in his twenty-one-foot Regulator. He lives in Manhattan but had rented a house in Montauk for the fall to be close to the fish. Thanks to him, I had a place to call home for the week. We found blues and albies decimating pods of bay anchovies, but we were unsuccessful with stripers. As the wind shifted to the southwest, it hummed across our lines at twenty-five knots, and we beat our way toward Diamond Cove Marina as conditions grew steadily worse. Soon we joined a procession of other boats entering the safety of the harbor channel. Later that night I received a message from Melnyk on my cell phone. "You should have been out there with me tonight, Dave. I'll bring the forty-four-pounder by in the morning."

He pulled into the driveway at 8:30 A.M. in his black full-size pickup truck. In the back, a big, broad tail poked out of a huge Igloo cooler. "Hop in," he said. "We're gonna go shove this fish up a few asses." Melnyk, who stands a healthy five foot eleven, had his head shaved to a fine fuzz, which accentuated his H-beam jaw. On one of his Popeye-like forearms was a tattoo of a striper turning on a bucktail.

Word had been spreading that Melnyk had been slacking off, or worse, that his luck had changed. He planned to quell those rumors with a little self-promotion. We drove from one fishing spot to the next, beckoning anglers to take a peek in the cooler. Most were unwilling, some were nasty, and others couldn't resist. "This burns them up," he said.

Our last stop was Camp Hero, where some SUVs were clustered together, wet suits hanging from tire racks. The owners, who had fished all night, were asleep inside—until Melnyk began pounding on their windows. One poor soul, who was wearing an eye mask, jumped so high that he bumped his head on the roof. He ripped off the mask to see Melnyk's smiling visage, then buried himself beneath his sleeping bag. "He can't help himself," said Melnyk. "He'll get up to see this fish." He did. And after we had sufficiently "shoved it up his ass," we continued on.

Melnyk discovered his unorthodox but deadly style of fishing by chance. Since the early 1960s, Montauk anglers have donned wet suits and like frogmen swim out to large rocks that, at low tide, still lay two or three feet beneath the ocean's surface. Once on these perches, the rock hoppers, as they're known, fished water that was unreachable by shore-bound surf casters. The results were extraordinary. One night in 1998, Melnyk was on Weakfish Rock, a large boulder that has a flat top the size of a kitchen table and sits some two hundred yards off the point, when a wave washed him off. It happens often, but this time Melnyk was fighting a thirty-pound striped bass, and it started towing him to sea. With no chance of hopping back on the rock, he decided to fight the fish in its element. His six-millimeter wet suit gave him plenty of buoyancy, and if he placed the rod between his legs and floated on his back "like an otter eating an abalone," he could actually put some leverage on the fish. Five minutes later he landed his prize. He was hooked.

III   III   III

After we debuted the big striper, Melnyk went to clean it and get ready for our first skishing expedition. I went home to my brother's place, sorted my gear, had lunch, and waited.

Melnyk arrived wearing his wet suit. With the top folded down to his waist, he resembled the barrel-chested prize-fighters of the late 1800s. After cleaning his fish, he had taken the head and set it on the front porch of a "friend's" house, complete with sunglasses and a half-smoked cigar poking from its mouth.

We drove onto the beach at Montauk Point and pulled on our gear. "I always feel like a gladiator putting this stuff on," he said. We zippered up our wet suits, put our dive belts around our waists (these held a knife and a set of pliers), grabbed our rods and flippers, and headed toward the water. On the way, we passed a fisherman in waders. "You ever worry about the bluefish?" he said, pointing to a darkened patch of water that had suddenly turned white with the froth of feeding blues.

"Nah," said Melnyk, "I'll just bite them back."

We waded in and pulled on our flippers. "We do everything on our back," said Melnyk. "Put the rod butt under your arm, lay back, and start kicking. I'll tell you when to stop." We pushed off and soon were gliding up and down the gentle swells.

Since this was purely a test to see if I could handle myself in the water, we didn't expect to catch many bass, but the blues were everywhere. When they began blitzing just ten yards away it sounded like a torrential downpour. To cast to them, I would give a quick flutter kick, which lifted me about waist-high. With my arms clear, I could deliver a good wing before I sank to shoulder level. I then put the rod butt between my legs and fished much like I would from land. I quickly hooked a few three-pound blues, which offered little fight on my ten-foot rod. Big striped bass, Melnyk told me, offer more of a challenge, especially when they get within fifteen feet. "Sometimes they spin you around, flip you over, pull you into the surf, and use their dorsal fins like bayonets." After I hauled in a few more blues and a schoolie striper, Melnyk pronounced me a natural, then called it a day. "We need to save our legs for tonight. It won't be too different from this afternoon, just colder, windier, and dark as hell." As we paddled in on our backs, Melnyk grabbed a moon jellyfish, a nonstinging variety shaped like the plastic lid of a cup, and took a bite out of it. "Not bad," he said, "tastes like watermelon." I began to wonder if I really wanted to follow this guy into the ocean at night.

III   III   III

I had Peter Benchley to thank for my fear of dangling my legs beneath the ocean's surface. I was an impressionable kid when *Jaws* opened, and for years afterward I took only lightning-fast dips in the river and at the beach. But for the most part I've outgrown this fear, damping it for good on a few late-night skinny-dipping excursions in the Atlantic with a blonde in my late teens. Or so I thought.

When Melnyk roared up the driveway to pick me up for our trip, flashing his lights in the darkness to let me know he had arrived, that familiar knot in my gut crawled toward my throat. Melnyk sensed my trepidation. "Don't worry about anything out there. There's nothing that can eat us . . . well, at least nothing has eaten me yet. And if it does, well, I'd much rather go like a man than lying in a bed. I'll drag myself out to the porch before I die in a bed." The macho prattle wasn't helping.

Within minutes we arrived at Ditch Plains, on Montauk's south side, the same spot where Melnyk had hammered the bass the previous night. He was stoked by the weather: flat calm, save for the swells coming in off the ocean. "You don't get ten days like this a year in Montauk," he said. "Perfect skishing conditions." We walked east on the rocky beach until we saw the light of someone else in the water. It was Melnyk's friend Bill, from the North Fork of Long Island. "I thought he might be out here tonight," said Melnyk.

Bill, who's known in Montauk as North Fork Willy, had hunted down Melnyk for a couple of weeks before he finally got him on the phone. "Told me he wanted to learn to skish, but I tried to scare him away with the danger lecture. Lots of guys are interested, but not many have the balls to do this. He was pretty persistent, so I took him out and gave him a few lessons. Turns out he's pretty damn good at it." Since skishermen turn on their headlamps only to unhook a fish or change tackle, we assumed Bill was into bass.

We entered the ocean in an area littered with boulders the size of compact cars. "Take your time until we start to punch through the waves. Then kick like hell. You don't want to get slammed on these rocks," said Melnyk.

Though the ocean wasn't rough, the swells coming in were at least three-footers. We hobbled through the field of rocks till we were waist-deep, then we put on our flippers. "All right," said Melnyk, "let's move." He quickly swam ahead of me, but I gained my momentum just as a wave rolled in. I rose to its crest as it began to break, but I slid through before it sent me shoreward. "Keep coming, we're not through yet," hollered Melnyk as another swell approached. This one I took a touch easier, cresting it before any white water appeared. "Nice work. Now, tell me this isn't fucking crazy!" said Melnyk. I couldn't disagree.

As we kicked, the swells flattened out and the sound of the surf diminished. "This is our natural state, man," said Melnyk. "This is just like being in the womb or floating in space." The womb I wasn't sure of, but space I could give him, especially since the night sky made a seamless transition to the sea, which was full of its own bioluminescent galaxies. Each kick stirred up microscopic plankton and larger moon jellyfish, causing the phosphorous in them to glow like cyalume sticks. Our entire lower bodies were bathed in a fluorescent yellow that slowly faded in our wake.

After plenty of shouting, we found Bill, who had a twenty-pound bass tethered by a stringer to his waist and was unhooking a nice weakfish. "The bite's on. I've landed six stripers already," he reported.

"I told you they'd be here," said Melnyk.

But they weren't, at least not for the rest of the night. The three of us spread out and began casting our eels. We worked an offshore current that took us east and then we swam into it for the next drift. Every so often Melnyk would lob a jellyfish at Bill or me to liven up the evening. He also told me about the time he and Bill got lost in the fog while skishing. "We were pretty turned around. We had to listen for the sound of breaking waves to find the beach. It took us about forty-five minutes," he said. "Then there was the after-noon in the electrical storm."

Bill's bass was still swimming on the stringer, and occasionally, when I drifted close, it bumped into me with its nose. Each time, I practically jumped out of my wet suit. "Sorry," Bill would say, "it's just my fish."

Bill's striper aside, I started to find skishing relaxing. But after two fishless hours, the three of us were pretty cold and called it quits. "All right, Dave, I told you getting out here was the toughest part, but I lied. Getting back to shore is the toughest part," said Melnyk.

As we neared the beach, we gathered ourselves and waited for a set of big waves to pass, then furiously kicked. When I felt my back skim over a rock, I knew we were close to shore. I could hear Melnyk yelling but couldn't make out what he was saying. Then a wave broke on me and I went under, taking a noseful of water, weeds wrapping around my body. I stood up, only to get crushed by another wave. When the water drained, I was knee-deep. I pulled my flippers off and stumbled to shore. "That wasn't too bad," said Melnyk. "You should try it with a forty-five-pounder tied to your waist."

⸻ ⸻ ⸻

Before there was Paul Melnyk, there was Jack Yee. The morning after my first skishing adventure, I took leave of Melnyk to spend time with Yee, one of Montauk's original rock hoppers. Yee took to the water in 1965 and is credited with taking the fringe fishing method to new levels. A fixture on the surf scene since 1963, Yee stands about five five, with a tanned face and teeth that look as if they were set by a nor'easter. I met him at 5:30 for his first beach run of the morning. The air in his truck was dank and fishy like a sea sponge. When we arrived at Gin Beach, it was still dark. We pulled out onto the sand and slowly started motoring toward the lighthouse. "This is my favorite time of day, because you can find plugs washed up on the shore that guys have broken off during the night." As dawn streaked across the sky, I took an inventory of his Jeep: a few bucktails and spoons and at least a dozen weathered plugs. Apparently the ocean washed up plenty of treasures. "Sometimes I even find cash," said Yee.

On one horrible morning he found the body of a good

friend washed up on the beach. The man had been fishing at night and suffered a heart attack. "I saw the body lying upside down, and when I turned it over I saw that it was my buddy Marvin," he said.

I asked Yee about his early days of rock hopping. Enraged by other surf casters who had no sense of beach etiquette, he decided to take to the water. "The first day, I wanted to go out to Weakfish Rock, but I approached it wrong," he said. "The tide was ripping out, and I was swimming and poling with my rod. I got my fingers on the rock, but I couldn't get on it. I swallowed half the ocean and everybody was looking at me. I eventually couldn't hold on and had to swim back in." On his next attempt he nailed the approach, and on his first cast with a pencil popper he hooked a thirty-five-pound bass. "That was my first big fish." By noon, a rumor had spread that Yee had whacked a fifty-pounder.

After that, Yee spent daylight hours drifting along the point finding large rocks that he could fish from at night. He eventually worked out a circuit of rocks he would drift to with the tide. While floating between rocks, Yee would smoke cigarettes he kept stashed in a waterproof camera-lens case. Since catch and release was not yet in vogue, Yee kept his fish on a stringer tied to his belt. "One night I caught a couple of blues and a bass that I had on a twelve-foot stringer. I decided to drift to a new rock, and when I got there I pulled up my stringer. A shark had eaten both of my bluefish. Only the heads were left. I didn't even feel a pull on my stringer." But sharks weren't Yee's only fear.

"Wet suits were real rubber back then, not neoprene," he told me, "so they were tough to get into. If you wore panty hose you could slip right into them. I was always worried I would wash up on the beach one day and the guy at the hospital would take off my suit and see Jack Yee in panty hose." The sea never washed up Yee, but a bad back has relegated him to stay onshore these days. "It's okay," he said, "I don't need to get wet to catch fish." Yee wears knee-high rubber boots and sweatpants. He's taken to life onshore with gusto.

These days, the sixty-five-year-old is known as the mayor of Montauk's surf scene. "If some guys on the beach have a problem, they search me out. They know I'm fair and people listen to me." They also come to Yee when they stick a treble hook in their hand. "I'm like the beach doctor. People knock on my door at three A.M. or come up to me in the middle of a blitz with a plug hanging from their finger, or their chest, or their back, and want me to remove it. The only place I won't touch is the eye. I've taken out at least one hundred hooks. Even the real doctors come to me. But they all get charged," he said. Yee requires his patients to let him pick one plug out of their surf bag. "I end up with some nice plugs." He learned his method—looping a piece of fishing line over the hook, pushing down on the shank, and yanking the line—from a promotional booklet on knots.

As we approached a fisherman standing on a stretch of beach paved with tiny stones, Yee rolled down his window. "Hey, get off of my rock," he yelled. The man turned around, smiled, and then backed up to the Jeep. "Hi, Jack. Kind of slow here last night. Couple of short bass on a needlefish."

"Yeah. The tide's heading in now. You should start to pick up some fish soon. Keep trying, wabbit."

"Wabbit" is Yee's term for anyone he doesn't consider a sharpie. The way I understand it, a wabbit often fishes unproductive water, but if he happens on a fish then he is mobbed by other wabbits. Wabbits and sharpies rarely mix.

As he rolled up the window, he told me, "I have no idea who half these guys are, but most of them know me." Before the window was all the way up, Yee turned to the man and yelled, "Don't you think it's time to get a job?"

Before long the marine radio in the Jeep crackled with life. "Whack 'em Jack, you out there?" When Yee picked up the mike, he made a sound that was a cross between the Road Runner's honk and a goose call. *"Whaank, whaank."* It's apparently his signature noise. The reports started coming in. In minutes Yee had a perfect picture of how the night had gone down and made a plan for the morning's fishing. "Playing the players," he said.

As we inched along the beach, Yee lowered the window. "Hey, I'm bringing the want ads down here Sunday," he yelled to a man in his seventies. We stopped at False Bar, and I started casting a yellow bottle plug. As the morning progressed, the spot drew a crowd and we left for lunch at a local diner.

When the waitress asked him how he wanted his burger cooked, he responded, with a comedian's perfect timing, "In the fire." She frowned.

"You want fries, too, Jack?"

"Yes. And tell the chef not to skimp on me again." She shook her head. Earlier in the week, Yee had complained that he'd received a measly portion of fries, and the chef had come out of the kitchen and dropped a bag of potatoes on his table.

After lunch we hit a few more spots. I caught fish at all of them. On the way home, Yee pulled into a gas station. "Check the oil, put air in the tires, and clean the windows," he yelled to a young kid at the pump. "And put in thirty-seven cents' worth of regular. Whaank, whaank."

III   III   III

Back at the house I was taking a nap when I heard Melnyk pull up on his motorcycle. Seconds later he came banging through the front door in his riding getup—black leather jacket, black boots, and a bucket helmet. He made an intimidating biker, especially since he wasn't happy. He'd received word that the previous night a rock hopper casting near Ditch Plains had landed a striper that bottomed out a fifty-pound hand scale. "I knew those fish were there. We should have stayed longer," he said. "Oh, well, we'll get them tonight. I'll pick you up at seven." As he walked out the door, he turned back and yelled, "It's not going to be pretty out there. The weather's turning a bit nasty, but you'll be fine."

Out in the water, a twenty-knot wind, which kicked up five-foot waves, and an angry bluefish on the end of my line worked in concert to give me the willies. After forty-five minutes I opted to head in. Melnyk indulged my fears. "I feel like a bit of a wimp," I said, as we kicked to shore. "Don't worry about it," he replied. "Face it, I'm just crazy."

In need of a stiff drink, I went to The Dock, where I ran into Captain Barry Kanavy. Among the bumper crop of fly-fishing guides working Montauk, Kanavy is the old salt. He spends much of his year fishing Long Island's Great South Bay, but trailers his boat to Montauk every fall. He's been guiding here for twelve years. I'd fished with him many times in the past and had him to thank for my largest striper on a fly rod, an eighteen-pounder. Though he's a strict conservationist these days, Kanavy was a commercial fisherman in his youth. He was shocked I got in the water. "You can't imagine some of the shit I've seen out here," he said.

"Sharks?"

"Yes. Big sharks."

I ordered a double, and the two of us hatched a plan for a morning outing. When I got home I washed off my wet suit and hung it up to dry, thankful my skishing adventures were over.

Kanavy and I met at the Montauk Lake Club marina and threw our gear in his flats skiff. Outside of the channel, we were socked in by fog. Land vanished, as did our chances of seeing another boat headed our way, but I couldn't imagine a more capable hand at the wheel. At age nineteen, Kanavy was a Master Helmsman on the USS *Essex*, an 872-foot aircraft carrier. One godforsaken night in the North Atlantic in 1970, he steered the ship through one of the worst storms ever to rock a navy vessel. Sixty-mile-per-hour winds had whipped the ocean into an endless range of seventy-foot waves. Walls of water rolled down the flight deck, which was fifty feet above the ocean. "I'd been at the helm for a good while when the captain of the ship came up and said, 'Son, do whatever you have to do to keep this ship moving ahead. You're not to leave unless you need to hit the head.' Then he asks me if there's anything I need, so I told him I could use a doughnut and some coffee. It's around three A.M., and he picks up the phone and calls the cook. We had a pot of coffee and a dozen fresh doughnuts in no time," he said. "I thought that was pretty cool."

Thirty-six hours later, Kanavy left the helm. During that time, pipes welded to the outer walkways had been ripped from the ship. In the engine room, layers of paint on the buckling and shuddering walls turned to dust, setting off fire alarms and panic. A few days later the ship limped into Boston for repairs.

With the fog hemming us in, Kanavy decided to head to Fort Pond Bay on Montauk's quieter north side. It's one of his favorite spots, but we knew our chances of landing a bunch of stripers were thin. Most of the fish in this area had followed the bait out to the point. We drifted among a sand-and-grass flat, dropping our flies in little pockets, talking about his children, jazz, writing, and the striper. "I think this fish means so much to me because I watched them disappear," he said. "I was on the water when catching a striped bass was almost impossible. One time I brought a bass to the marina and the guys at the dock didn't know what it was. Can you imagine that now?"

New York: Urban Angling

Fishing the New York Harbor environs is part history lesson, part tourist attraction, and part urban adventure. It seemed just the place to introduce the first and only guest of my trip, my nine-year-old nephew Robert, to the pleasures of striped bass.

Robert is my oldest brother's son. He looks like he jumped straight out of a Rockwell painting—blond hair, blue eyes, freckles, and two front teeth yet to be reined in by braces. He also has a large helping of self-confidence. His brothers and sisters call him HB, short for Humongous Big-head, after a character in *Harry Potter*.

HB lives in South Carolina, surrounded by a sea of pine trees, cattle pastures, and farm ponds. I couldn't wait to see his face when he set the hook on a large striper headed for the safety of deeper water.

The night before Robert arrived I drove into New York City, home for the first time in almost two months. After sorting through mail, I pulled my books on the harbor from the shelf. I figured I'd brush up on some history to possibly teach Robert a few things while we fished.

III   III   III

A sprawling estuary that includes a fjord (the lower Hudson River) and a tidal strait (the East River), New York Harbor offers a plethora of habitats for different types of fish. It didn't take long for the settlers of New Amsterdam (the name was changed to New York when the British defeated the Dutch in 1664) to do more than dip their toes in the harbor. By the early 1800s, well-dressed anglers were making quite a sport of hauling large sharks, some up to twelve feet, from the East River with a hook tied to rope and chain. Even lobsters were said to reach six feet.

The East River's Hell Gate, a treacherous passageway for sailing ships, was lined with reefs and guarded by a screaming ten-knot current. In his book *Heartbeats in the Muck,*

John Waldman reported that one out of every fifty vessels making the run was sunk. The combination of current and structure also made it a tremendous place to land striped bass, and a fishery flourished on the banks. Sportsmen could rent boats for $1.50 a day to fight the stripers and the current. Waldman writes, "Guides manned the oars as their patrons trolled; anglers fishing alone held their lines in their mouths as they rowed. To withstand strikes, solo anglers had to have their original teeth." In the late 1800s, the reefs of Hell Gate were blown to bits by countless dynamite charges to make for safer shipping lanes.

As the population boomed on the island, New York Harbor soon resembled a giant marine septic tank. In 1910, six hundred million gallons of raw sewage was spit out into the harbor environment each day. Children who went swimming in the East River came home coated with a black sludge. In some places the waste piled ten feet deep on the harbor floor. Out at sea, a plume of sewage could be seen leaving the river on an outgoing tide. Marine life lucky enough to have tails simply swam out of the harbor and didn't return; sessile creatures, like oysters and clams, suffocated. The most eloquent chronicler of New York Harbor, Joseph Mitchell, wrote in 1960, "The bulk of the water in New York Harbor is oily, dirty, and germy. Men on the mud suckers, the big harbor dredges, like to say you could bottle it and sell it for poison."

There was little change until the passage of the Clean

Water Act in 1972. Nowadays there are swimming races held in the harbor. And while sheepshead have yet to return to Sheepshead Bay, and six-foot lobsters are a gourmand's pipe dream, the swimmers share the water with hundreds of thousands of striped bass.

In the fall, stripers pour out of Long Island Sound into the East River and eventually out of the mouth of New York Harbor on their way south. In the ocean they join ranks with the fish coursing down the Atlantic. The waters of the harbor, however, are more than a simple conduit. Stripers linger, feeding in the fierce currents as they have for thousands of years. Their favorite feeding zones include the helicopter pad behind the United Nations Building and the rips at Lady Liberty's massive green feet.

||| ||| |||

Robert arrived at LaGuardia Airport on a U.S. Air flight. After everyone else had left the plane, he came running out of the gate with a flight attendant in tow. As he hugged me, I signed off on some papers and we headed for the baggage claim. "When can we go wadering?" asked Robert.

"Wadering?"

"Yeah, fishing in my new waders," he replied. I'd had his father buy waders as a backup option in case it was too rough to get out in boats.

"We may do some before the end of the week, but first we have some trips planned with the best guides in New York City. We're going to be fishing from their boats. I've heard the action is really hot," I said, trying desperately to shift Robert's focus.

"Oh," he said. "Can we go wadering instead?"

The next morning we woke up at 3:45. We had to meet Captain John McMurray at Beach Channel Marina near Far Rockaway, Queens. Once a vibrant beach town, Far Rockaway sits on the western tip of Long Island, as far from Montauk as you can physically get. Nowadays, its quaint bungalows sit tight in the shadows of gritty apartment buildings.

In the car, Robert fell asleep before we were two blocks from my apartment. When we arrived at the marina an hour later, I had to practically stand him up before he would wake up. "We're already here?" he asked, rubbing the sleep from his eyes. It was still dark and cold—thirty-five degrees. The parking lot, which was capped by a large chain-link fence, was starting to fill with boats that had been hauled for the winter. When we met McMurray, Robert looked him directly in the eye and shook his hand. I could tell my brother had prepared him for the trip.

Robert admired McMurray instantly. I think it had something to do with his beard, à la Grizzly Adams, and his jumpsuit. Robert has worn jumpsuits to go hunting and fishing since he was three years old. He's a connoisseur. McMurray's jumpsuit was a relic from his Coast Guard days at nearby Rockaway Station. It was a bright red number as thick as a down jacket.

McMurray wanted to get on the water before sunrise, so we idled out of the marina in darkness. By noon McMurray would be at work at the Norcross Wildlife Foundation in the city, where he decides what environmental causes deserve the group's money. It's a tough schedule, but like many guides, McMurray believed the fall to be a reward for some of the doldrums of summer. "Sleep is a precious commodity during the run, but I've got all winter to sleep," he said.

Just outside of Jamaica Bay, with the sun rising in our wake, we encountered what the writer Denis Johnson called the "sour electrical" smell of the World Trade Center ruins. McMurray shook his head. Robert wrinkled his nose and looked up at me. I pointed to the tip of Manhattan, where a cloud still emanated from the site. "Is that Ground Zero?" Robert asked. I nodded. The image of destruction had become iconic.

Since September 11, the waters of New York Harbor north of the Verrazano-Narrows Bridge had been off limits to pleasure craft. The closure sealed a productive fishery from guides who mentioned working Wall Street in the context of tossing Clouser minnows toward the bases of skyscrapers. But even with many honey holes being patrolled by small gunships, there were miles of fishable water in the harbor vicinity.

When we reached the Breezy Point jetty, McMurray eased up on the throttle. The birds were resting on the water; a few flew reconnaisance high in the sky. Not much was happening, but sitting at idle, we had a chance to warm up. The jetty sits on the northern side of the entrance to New York Harbor. Offshore, as the day grew lighter, we watched huge ships materialize. They were waiting to be escorted into the harbor by tugs. Behind us, traffic on the Verrazano-Narrows Bridge was falling into a typical stop-and-go pattern. I had Robert make practice casts with a spinning rod to warm up. I watched like a proud father as he sent his lure zinging into the morning. "That's better than most adults," said McMurray.

As soon as the tide turned, McMurray expected the fish to rise, and they did. Just off the beach, birds started squawking, and we spotted a school of bass harassing peanut bunker. Robert's eyes grew wide as we approached, and he readied his rod. I knew he would hook a fish if he landed his lure near the melee. His cast, borne of a rush of adrenaline, was a lame duck. The lure landed five feet from the boat, and the fish sounded before he could reload.

On the next pass, Robert dropped his Fin-s-Fish in the zone and hooked up. Seizing the moment, I grabbed a fly rod and made a few false casts before delivering my fly. When it landed short of the fish, Robert looked up from his battle and said, "That's not far enough, Dave."

Robert's first striper was about twenty-two inches long, about three inches longer than the one I eventually hooked. He was back on top.

As the day progressed, I watched McMurray grow larger than life in Robert's eyes. McMurray sealed his status with his story about Floater Week. It seems every April, the warming waters encourage human bodies that have found their way into the harbor over the winter to float to the surface. Retrieving the waterlogged corpses, most of which are the results of murders or suicides, is the Coast Guard's duty, and McMurray landed plenty during his hitch. "What was it like?" asked Robert.

"It wasn't too fun. But one thing I knew was that when the bodies started floating, the stripers weren't far behind."

Robert smiled. I imagined him in the nights to come dreaming about a school of striped bass nosing a floater out of their territory. He was getting the true New York experience.

Later in the day we moved to the Coney Island Rip. Onshore, a huge Ferris wheel and the parachute drop sat idle. Robert hooked a ten-pound bluefish on the first drift. I followed with a nice striper. Robert pointed to a boat next to us: "He's got one, too."

McMurray shook his head. "Let's see what he does with that fish, Robert." The guy was fishing a bucktail with a hand line, essentially a thick cord tied to a piece of heavy monofilament. He had one in each hand. As the boat trolled through the rip, he would snap his arm back as if yanking on a dog leash. When a striper hit, he tied the other line to a cleat and hauled his fish in hand over hand. When the fish appeared just behind the boat, the angler swung it aboard and dropped it into the cooler. It was an illegal-size striped bass. "I've been watching him all morning. That's the seventh striper I've seen him keep—six over the limit. He's out here all the time," said McMurray.

Poachers try to slip in between the cracks wherever there's an opening, and New York Harbor provides plenty of crevices. McMurray has been fighting this battle since he was a boarding officer at the Coast Guard's Rockaway Station in 1995. It wasn't the Coast Guard's responsibility (that belongs to the Department of Environmental Conservation), but McMurray "made it a priority to jam those guys up." He once busted the same poacher we were watching. "I made him hand over all of his knives before I let my crew board him. He had more than two hundred pounds of illegal fish. They let him off with a fine, and he was out here the next week." It was a bittersweet victory, especially since they were so difficult to come by.

The poachers keep their illegal fish in weighted burlap

sacks, which they simply toss overboard at any sign of trouble. It's speculated that some boats make three trips a day, landing a total of as much as five hundred pounds of illegal fish per boat. McMurray figures they take their catch to Chinatown and sell them off the truck. Not only is this against the law, it's dangerous. The Department of Health warns that women of childbearing age and young children should not eat fish from the harbor environs, due to high PCB levels. Anyone else should eat those fish no more than once a month.

McMurray had recently received a tip that the poachers had tie-ins with organized crime. It was disconcerting news, but he vowed to stop them. "Those guys are stealing our resource from under our noses," he said.

Since September 11, every available law enforcement agency was concentrating on security, giving poachers free rein. "Can't we do something, you guys?" Robert asked.

"Trust me, Robert, if there was anything I could do, I would do it," said McMurray. "I'll take some photos of the boat and send them to the authorities." On the way home, we wolfed down sandwiches. It was time for McMurray to start his day job.

||| ||| |||

Back in the truck, Robert and I were bogged down in traffic on Cross Bay Boulevard. After Robert called his father, he promptly fell asleep for the remaining hour and twenty minutes of the trip. At my apartment, I turned on the TV for Robert and lay down in bed. Fifteen minutes later I woke up to his hand buzzing back and forth just above my nose. I could hear him giggling. "The way I look at it, Robert, you racked up two hours of sleep in the car. Now it's my turn," I pleaded.

"Okay, okay. I'll let you sleep some more." Twenty minutes later he launched a tennis ball from the hallway that landed directly on my stomach. That night at dinner, Robert informed me that the Chinese food we were eating tasted a lot different than what he was accustomed to in South Carolina. "That's not necessarily a bad thing," I told him.

"Tastes bad to me," he said.

The next day, Thursday, we were up at 4:00 A.M. We had a trip with Frank Crescitelli, possibly the most obsessed fisherman I had met on my trip. Crescitelli grew up on Staten Island, where he now lives with his family. He runs fishing trips all week but keeps Thursdays open for his own enjoyment, which means more fishing. Crescitelli has made it his mission to awaken New Yorkers to the fabulous angling that lies outside their back doors. Until September 11 he had often picked up clients at the World Financial Center's dock before and after work. On that fateful day he was on the water. "We saw the sun flash off the wings of the second plane," he said. Crescitelli rushed to his marina, filled his tank, and went screaming toward the destruction. "I figured

I could help ferry the injured, but the Coast Guard wanted me to assist with security. I didn't even have a gun."

We began the morning chasing pods of stripers off Sandy Hook, New Jersey, a literal hook of sand that extends deep into the harbor's mouth. After thirty minutes I left the bow to check on Robert. When I reached the stern, I saw that a school of stripers had surrounded the boat, their tails thumping off the fiberglass as they madly pursued bay anchovies. As I turned to grab my own rod, Robert reared back and let fly, his lure snagging my wool hat and yanking it off my head. When I turned around, it was floating near the boat. The fish were boiling all around us. "Better reel that one in, Robert," said Crescitelli. "It looks like a nice-sized one. Tough to eat, though. Kind of stringy."

Then we got the call from McMurray. As is normally the case these days, the guides communicated with cell phones—announce the location of fish over the radio and you'll draw a crowd in minutes. The fish were stacked up at the Coney Island Rip again. Crescitelli slammed down the throttle and the boat started dancing across the chop. We arrived a few minutes later.

The rip looked like a formidable place to catch a fish. The current swept over the shallow bar and forced waves to stand head-high before they broke with a whoosh. McMurray was about three hundred yards away. The fish, he had told us, were feeding in the white water after the waves broke.

I picked up a rod with a popper, planning to show Robert how to work it. When it hit the water, I gave it a tug and it chugged toward us. Before I could make another twitch, a massive striper swatted the lure with its tail, most likely in an attempt to wound it. The lure, now looking like a kite, went flying in the air. I yelled for Robert to make a cast with the closest rod. I chugged again, and this time the lure disappeared in a giant boil. From the bow, Crescitelli let loose with a howl as he hooked up with his fly rod. Robert connected next. A triple-header. With all fish released we motored back to the rip, where the swells seemed to have grown in size. We all made a cast. Another triple-header. "This is the hottest day of the fall," yelled Crescitelli. All of the fish were between ten and fifteen pounds.

A few drifts—and many more stripers—later we heard Crescitelli yell again, this time his voice hoarse with urgency. "Oh crap, hold on, guys." When I glanced forward, I knew why. Just ahead of the bow, a swell seemed to stand on end. It looked to be about four feet above the pulpit, where Crescitelli was now crouched, clutching the bow rail. A feathering of white water gurgled on the rogue wave's crest. I grabbed Robert as the bow rose straight up, then dropped with a thud as the wave zoomed under us. As it passed the stern, Robert and I were showered with spray. "Here comes another," yelled Crescitelli. Mercifully, this wave was a bit smaller. The boat rose and fell. Crescitelli let out an exhilarating scream. Robert's hand grasped my pants, his knuckles as white as Wonder Bread. "Okay, Dave. I'm scared now," he said.

"Don't worry. We're fine."

"Dave?"

"Yes?"

"Can we go wadering tomorrow?"

III   III   III

The inaugural wadering trip took place in the back of Manhasset Bay on Long Island. I knew of a sandy bank that offered little threat of breaking waves and a chance for stripers. At the truck, Robert was the picture of pride, strutting about in his new waders. I cinched him up with a belt, showed him how to rest the rod on his shoulder, and snapped his suspenders for fun. Minutes later, thick in some marsh grass on our way to the beach, his enthusiasm waned when he stepped in a hole and went down. He rose dripping wet, eyeing me as if I'd orchestrated the fall. "Welcome to the joys of wadering," I told him.

At the beach he waded in slowly. With the water at just below knee level, he began to do a series of squats. "You all right?" I asked.

"Yeah. This just feels real weird. There's like suction on my legs."

"That's normal," I told him. "Hold my hand and we'll go deeper so you can get used to it."

"Nah. I think I'll just fish here," he said, backing up to ankle depth.

Though I hooked one bluefish that Robert reeled in, the action was slow. Before long we were skipping rocks and having casting contests.

That afternoon I took Robert to the airport for his return flight. When I called his house to make sure he made it my brother answered the phone. "Robert can't stop talking about *wadering*," he said. "Sounds like you guys had fun."

III   III   III

The following day I took the Ninety-sixth Street crosstown bus to First Avenue, where I hopped out into a spitting rain on my way to the East River. While I had often fished the Hudson, I had never explored the waters of the East River. Unlike the Hudson, which is buffered from the city for much of its length by Riverside Park, the East River butts up against the hard edge of Manhattan. Take a big backcast and you'll hook a taxi or put your sinker through a factory window. Fishermen here have no qualms about using their rod as a weapon if threatened by crime. It may not be skishing, but you'd be hard-pressed not to call it extreme.

Once I made it across the FDR Drive I headed north, with the East River on my immediate right. I saw my first condom floating near 103rd Street. At 110th Street I came upon fishermen. Three men stood watching their rods; they each had two. Rigged to the top of each rod was a bell that alerted them to a nibble if they weren't watching. The rods were leashed to the steel fence by an assortment of ropes, wires, and even a plastic bag. On a small patch of ground, the men had put up a blue tarp that was strung between a lamppost and a scraggly tree. Underneath it were two chairs, a couple of backpacks, some bags of bait, a small radio, and a nice bluefish. A few feet away, cars and trucks barreled down the FDR, swerving, honking, and spewing exhaust.

I approached a man wearing a yellow poncho and asked him about his luck. "Just a small blue," he told me and then went to tinker with his rods. I had more luck with the younger guy in the bunch. His name was Ponci. He was from Puerto Rico but had lived in the city for twenty years. Ponci showed me his rod, an eight-foot Ugly Stick with a Penn spinning reel. "It was a present from my wife for Valentine's Day," he said.

Ponci informed me that the man in the yellow poncho was the pro. His name was Jose, and he fished here every day. Just last week he'd landed a thirty-six-inch striped bass. Ponci's personal best was a thirty-two-inch striper that he landed from the point of Randall's Island, directly across from where we were standing. The riprap along its shores makes it a prime spot to catch stripers. Unfortunately, the rocks have a habit of eating expensive tackle, and Ponci and Jose rarely risk it.

Once Ponci had lowered a small net to trap bait and caught a sea horse. "I couldn't believe it, man. Its face looked just like a horse's. I put it in a bottle of water and watched it swim around for about an hour. Everybody wanted to see it, but that face just made me sad. I mean, it was a tiny horse face. I let it go and watched it swim back to the bottom."

Just then the bell on Jose's rod jingled. "Jose, Jose, your rod," yelled Ponci.

Jose, who was cutting some bait with his back to the rod, didn't turn his head. "*Pequeño. Muy pequeño,*" he said. The rod stopped dancing. It reminded me of the great jazz pianist Art Tatum. Though legally blind, Tatum could name the type of coin dropped on a wooden table by the pitch of the sound it made when it landed.

"Jose's the man. He's the man," said Ponci.

Before I left, my friendship with Ponci seemed to loosen up Jose. He told me about his big striper and showed me the giant treble hook he lowered over the wall to hoist the fish up. "Fishing is my work and my love," he told me, making a fist with his hand and tapping it on his heart.

"People must think you guys are crazy, fishing out here in the rain just three feet from the road," I said.

Ponci piped up from behind me. "Let them see how beautiful a rod looks when it's jumping and bending with a big striper on it and they'll know why we stand out here in the rain."

New Jersey: A Fish Story

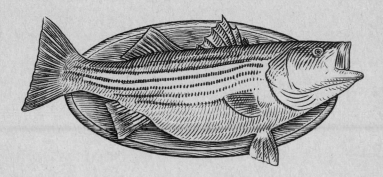

Twelve thousand years ago, a massive glacier that had crept south from Labrador lost its oomph near the mouth of the Hudson River. Where it stopped, an imaginary line was drawn east to west. North of it would become classic striper water, south of it average striper water. As the glacier retreated, it gouged out deep harbors and left behind rocky shorelines, textbook habitat for striped bass. South of the final advance, much of the coast would become exceptional for its uniformity—a long, thin stretch of sand that's known as the Jersey Shore. In early November, I crossed the line on a still day at 6 A.M.

On the New Jersey Turnpike, the sun rose to my left, coloring the cloud from the World Trade Center site almost peach. The hole in the skyline was raw and open. I had stayed in the city for an extra week. The fall weather had been beaten back by record warmth, and the migration had stalled. I had enjoyed the rest and time with friends, but I couldn't wait to get back on the road. The turnpike—lined with factories burping plumes of smoke and giant gas tanks—was not what I had in mind. Though I was expected in Seaside Park by lunch, I took one of the first exits I could and headed east, looking for a road that ran along the shoreline. When I found it I headed south again, stopping when I saw some bird activity offshore.

As I readied my gear, a trio of elderly women walked past on the boardwalk. From a snatch of their conversation I learned that they were snowbirds, soon heading for the tropical delights of winter in Florida. Off the beach, the terns had settled down and now drifted on the swells. On the end of nearby jetty, I tied on a yellow Bomber and made a cast. I couldn't beg a strike for the next forty-five minutes.

I saw some fish rise in front of a jetty two hundred yards down the beach and went jogging to them. On the jetty, a big man in a baseball cap, large sunglasses, jeans, and rubber fishing boots let me know I was too late. "I used to run after the fish, too," he told me. "But now I don't. They'll show up again."

Stuart Wilk, it turns out, knew quite a bit about the striped bass. A retired thirty-seven-year veteran of the National Marine Fisheries Service, he was heavily involved with the Emergency Striped Bass Study in the early 1980s. Like a farmer surveying his fields, Wilk scanned the ocean. Unlike many Jersey fishermen, who felt they had begun to see less fish recently, Wilk was content with the present state of the striper. "We have a healthy population, and in five years we're going to see some big fish. I'm already seeing more twenty- and thirty-pound stripers than in years past," said Wilk, who lives just ten minutes from the shore and had spent a better part of the fall season fishing from its jetties.

He did voice concern over New Jersey's alarming trend of replenishing every inch of coast with sand sucked from the sea bottom. It wasn't the first time I'd heard that complaint from an angler. The state is the ne plus ultra when it comes to the hopeless task of stemming the ocean's rise. Since 1994, tens of millions of cubic yards of sand have been dumped on Jersey beaches. The sand is pumped to the beach from offshore "borrow pits" via a large ship and a system of massive pipes. Though much of it is garden-variety beach sand, some comes ashore as green, stinking mud. Onshore, bulldozers level the sand. In some places it may pile ten feet high. Often entire jetties are buried, as is any nearshore habitat, such as reefs. A once-varied sea bottom becomes a lunar landscape. To the delight of fish and fishermen, the sand, after sunning itself for about three years, migrates back to sea. It's become the government's equivalent of digging a proverbial hole in the sand.

Wilk pointed to a flume near the base of the jetty. Water trickled into the surf. The flume connected a nearby pond to the sea. Before there were houses and seawalls, it had been a small creek. Each spring, its water turned as dark blue as a bruise when thousands of herring rushed its flow to spawn in the pond. Each fall, the product of that mad dash once again tinted the waters and new life returned to the sea. When it did, the nostrils of migrating striped bass followed the oily scent of the herring to shore. The herring had survived the transformation from creek to flume, but now sand clogged the link.

Wilk said the sand had been pumped onto beaches south of the jetty and had migrated with the offshore current. It gathered in a large sandbar at the mouth of the flume. Any fish making the run risked getting stranded or simply picked up by seagulls hopping around the shallow cut. "At low tide, the herring don't have a chance," he said.

"What lake is that?" I asked, still not sure even where I was.

"Deal Lake. This is the Eighth Avenue Jetty."

I was shocked. I had stumbled upon Asbury Park, New Jersey.

In 1888 the town's founder, James Adam Bradley, created the state's first fishing club. Its headquarters were on a pier that stood just south of the Eighth Avenue Jetty. Bradley donated a bell and four rocking chairs to the cause. When a striped bass hits the planks, the bell was rung—one chime per pound. The Asbury Park Fishing Club is still thriving. Asbury Park is not.

The town boomed for nearly a hundred years, gaining a national reputation as a music hotbed and a premier shore destination. In 1969, however, the town's death knell sounded when race riots ripped through the area. The scourge of drugs followed the unrest. By the 1980s, vials of crack were easier to obtain than saltwater taffy. Lately, famous son Bruce Springsteen, who spent his youth in Asbury Park honing his skills, has been leading a revival effort. It hasn't been easy. Many buildings on the boardwalk are boarded up, and a half-built luxury high-rise mars the beachfront. Residents call it "Beirut by the Shore."

While I had found a historical fishing spot, I hadn't found a hot spot. For the rest of the morning the fish stayed out of casting range. The surface action, however, was encouraging. The stripers were making a move.

I spent the next two days in Seaside Park, a small town that sits at the entrance to Island Beach State Park. The park is an untouched slip of land known as the center of striped bass action in the state. Bob Popovics, a master fly-tier, and Fred Bogue, a fishermen so passionate about stripers he had two of them on his wedding cake, were my guides. Though we fished hard, we only landed a few small bass. The guys in the boats reported big schools of stripers hanging about a half mile offshore.

||| ||| |||

I pulled into Atlantic City looking not for fish but for a fisherman. I had heard stories about him since the day I hit the road. Most anglers I met, though they didn't know the man, despised him. I'd been told he was a drifter, a cheat, a junkie, and an alcoholic—all of it unsubstantiated. No one could deny, though, that Albert McReynolds held the current world record for striped bass. Through a search of newspaper databases and interviews with the few people who claimed McReynolds as a friend I had pieced together his story before I arrived. The more I learned the more bizarre it became. It all started on September 21, 1982, the night McReynolds hooked the striper of the century.

McReynolds, who lived in Atlantic City at the time, was casting from the Vermont Jetty, a narrow rock bulkhead that extends roughly twenty-five yards into the Atlantic. As striper spots go, it's rather nondescript, but McReynolds and a group of fishing buddies who called themselves the Wet Ghosts had a fair amount of success on the jetty that fall.

A nor'easter had barreled into the coast, sending sheets of rain into boardwalk casinos and massive eight-foot swells lumbering to shore. The majority of the jetty was awash. But McReynolds and his partner, Pat Erdman, a local fireman, took root. In the water, schools of mullet rushed along the length of the jetty. Each time a particularly large wave broke on the rocks, dozens of the five-inch-long baitfish were left flopping at the fishermen's feet. McReynolds knew that striped bass had the mullet pinned to the jetty and were gorging on them beneath the chaos of white water. In an instant, Erdman hooked up with the first of many nice stripers.

None of them would compare to the giant that swiped McReynolds's lure, a five-and-a-half-inch black-and-silver Rebel swimming plug, at 10:00 P.M. On her first run, the striper peeled two hundred yards of twenty-pound test line off McReynolds's reel.

McReynolds fought the cow for an hour and twenty minutes before she finally surfaced on her side. Now he had to step into a cauldron of raging water to land her. It's a tricky endeavor when the weather is calm, requiring the agility of a sandpiper. When the weather is sloppy, it can be deadly.

After a botched attempt to land her with a hand gaff, McReynolds was swept into the water. When he surfaced, the fish was floating right in front of him. McReynolds thrust his fist under the fish's gill plate and out her mouth.

With Erdman's assistance, McReynolds lugged his prize onto the jetty. "I got the fish up to the rocks, and I more or less collapsed on top of her," he later told the Associated Press.

There would be no tackle shops open for weighing the striper until morning, so the two men put a wet blanket over the fish and strapped it to the roof of Erdman's car. With the striper secure, they continued fishing. As dawn approached, the two knew Corky Campbell would soon be readying his tackle shop for the day's business, so they headed there.

At Campbell's Marine Bait and Tackle shop, no sooner had McReynolds hoisted the fish on the scale than the phone began ringing at a rate of twenty-five times an hour. It weighed seventy-eight pounds, eight ounces, eclipsing the former record by two and a half pounds. Friends poured cans of Budweiser over McReynolds's head, and soon a party was raging at the shop. By the afternoon, fifteen hundred people had visited the shop to gawk at the fish. Some, like worshipers at the Wailing Wall, touched the fish reverentially. Many speculated about its live weight—its weight immediately after it was removed from the surf. Like a water balloon with a slow leak, a fish on land loses weight fairly rapidly. Most agreed that the striper would have pushed eighty-two or eighty-three pounds if it had been weighed just after it was landed. Then Nelson Bryant from the *New York Times* called to tell McReynolds that he could be eligible to win $250,000 in a sweepstakes that Abu-Garcia, a tackle manufacturer, was running. The rules stated that the first person to catch a world-record striper, salmon, trout, or largemouth bass in 1982 would win the money. Even Campbell would receive $25,000 for weighing in the fish.

A few days later, Edward Keating, a sports agent who worked with Dick Butkus, Jackie Robinson, and Arnold Palmer, also phoned to offer his services. There was even a call from a lawyer representing a posh hunting lodge. The owner of the lodge was willing to pay $250,000 for the skin mount of the fish. The striper was insured for $100,000 and locked in Campbell's bait freezer. But before McReynolds could claim the record, the International Game Fish Association (IGFA) had to verify that the striper was caught in compliance with its rules. It's a common practice, but rumors about McReynolds's catch were already swirling.

An anonymous letter had been sent to the IGFA claiming that Erdman and McReynolds had found the fish floating in the surf. Others suggested that McReynolds, who was a former commercial fisherman, had been given the fish by one of his cohorts who had caught it in a net. Surf fishermen on the Jersey coast who had been out the night of the catch argued that there was no way McReynolds could have set foot on the jetty, much less landed a fish from it. Big swells, they said, would have washed him right off. The whole story just seemed, well, fishy.

The investigation included a dissection of the fish, a

review of photographs taken at the weigh-in, and intense interviews of those close to McReynolds. Though the IGFA didn't require it, McReynolds and Campbell brought the fish to have it x-rayed to prove it had not been stuffed with extra weight. Four months later, the IGFA certified McReynolds's catch.

Soon afterward, McReynolds was presented with the $250,000 at an awards dinner at the Explorers Club. He also received some endorsement money from the tackle companies whose products he was using. The fish's worth was estimated at nearly $3,200 a pound. *The Guinness Book of World Records* listed it as the world's most valuable game fish.

Wearing the laurel wreath of world-record holder, however, felt more like a crown of thorns to McReynolds, who had lived day-to-day in motel rooms with his family before the catch. Hate mail from anglers flooded his mailbox. He became mired in squabbles over money. Everyone wanted a piece of the action, but Keating was handling the funds. McReynolds moved his family into a dream house on Cape Cod, but the cold, damp climate exacerbated his wife's asthma and they left. They spent a stint in Hawaii but eventually returned to Atlantic City. There was no warm homecoming. McReynolds couldn't even find work. He was considered an outcast.

Fishing was the only constant in McReynolds's life, so he loaded his family in the car and began an angling odyssey that continues to this day. During the winter and spring, they fish in the south. In the summer and fall, they move. Along the way, they sell bait to tackle shops to make ends meet. His daughter found a way out of the backseat by marrying a guy from the Midwest. McReynolds's wife and two sons still travel.

III   III   III

I'd heard McReynolds spent part of November in Atlantic City fishing a local tournament and selling bait. But locating him, I quickly found out, would be as difficult as breaking his record. No one had a phone number for the family. Short of running into McReynolds, there was really no way to reach him.

The tournament had ended the day before I arrived. The McReynolds family had captured nine awards—a record— but the word was that they might have already skipped town. In a dingy motel room across from Trump's Taj Mahal, I ripped a page from the phone book and started calling local tackle shops. They all knew McReynolds. It seems he makes a circuit of the shops, peddling bait and talking fishing. One frank individual told me, "You could go deaf before he shuts up." When I asked another guy where I might find McReynolds if he was in town, he said, "Try a bar," then hung up. Nearly twenty years had passed, but the animosity was still there.

I got a better reception when I called the Absecon Bay

Sportsman Center in Absecon, New Jersey, just outside of Atlantic City. I was told they had the Penn spinning reel McReynolds used to catch the record on display and that McReynolds had recently sold some bait to the shop. I hung up the phone and drove to Absecon.

The shop and marina sat snug on Absecon Creek. Outside, bait tanks gurgled and sputtered and a brace of freshly gunned ducks rested in a clump on the cleaning table. I arrived just in time for the Friday-afternoon rush. The line at the register was four deep, and other customers milled around the massive racks of tackle. Two Labs lounged in the main aisle, unmoved by the crowd.

When I asked about the reel, the young kid I'd spoken to on the phone pointed to a glass case by the cash register. "You're the one who called. It's on the right." But it wasn't. It turned out that McReynolds had taken the reel recently to sell it. "Sorry. I didn't know it was gone." I did learn that the owner, Dave Showell, an experienced fishing and hunting guide, often spoke with McReynolds when he was in town and respected his angling talents. He taped my card to the register and promised to have McReynolds call me if he showed.

While waiting on McReynolds, I made two goals: to find a mount of the record fish and to catch a striper from the Vermont Jetty.

The first wasn't as easy as I had hoped. Though several mounts were made, only one remained in the Atlantic City area. It seems the foam fish became a hot potato of sorts, bouncing from one shop to the next. Rumor had it that McReynolds had shopped it around to the highest bidder. I eventually discovered the mount hung in Gene's Beach Bar in Brigantine, New Jersey, just over the bridge from Atlantic City.

The Beach Bar is a low-slung structure near the ocean. The striper was hanging by the door in a backlit glass case. It struck a classic fish pose, bent in a slight U shape, mouth agape, fins erect. And while I'd heard how big this fish was, I wasn't prepared for it up close. You could have almost stuffed a volleyball in its mouth, and its girth, thirty-four inches, was as big as my own waist. Surrounding the fish were a few faded photos of McReynolds hoisting it up by the gill plate. The striper was nearly as long as McReynolds was tall.

The big fish was part of a class of striped bass born in the late 1950s. They had changed the landscape of the sport. Between 1980 and 1982, anglers captured four seventy-pound-plus fish. These stripers were twenty-five or more years old and had achieved their grand size through a combination of variables that no doubt included genetics, plenty of forage, and lots of luck.

The first to fall was John Baldino's seventy-one-pounder caught off of Norwalk, Connecticut, in July of 1980. Almost exactly a year later, Bob Rocchetta hammered a seventy-six-pound striper on a live eel while fishing the rips around

Montauk. Tony Stetzko's seventy-three-pound striper, landed in November of 1981, was next. Then came McReynolds's catch.

I went to the bar and ordered a beer. The patrons, mainly blue-collar types in construction boots and baseball hats, were just coming from work. Classic rock blared from the jukebox. I asked the bartender if McReynolds ever showed up at the bar. She shrugged her shoulders. "What type of fish is that, anyway?"

"A striped bass," I told her. "A very big striped bass." She nodded her head ambivalently, spun around, and tossed a rolled-up napkin in the trash.

On the way to the Beach Bar I'd gotten lost and stopped to ask directions at a tackle shop. "If you need any motivation to go out there and fish all night when it's cold and nasty, then go see that fish," a blond-headed kid with a ponytail told me. "If that doesn't fire you up, nothing will." He was right. After I finished my beer, I went to catch my striper off the Vermont Avenue Jetty.

ııı   ııı   ııı

Characters abound on the Atlantic City Boardwalk, especially at midnight. High rollers leave their seats at the baccarat table for some fresh ocean air, prostitutes troll the strip looking for the high rollers, drifters wander through the crowds wondering about their next dollar, and vendors hawk trinkets and corn dogs. Still, as I walked along in my waders, my ten-foot surf rod resting on my shoulder, I caused a bit of a stir. A group of teenagers smoking cigarettes made sure I heard them laughing and a man with a snake tattoo slithering down his forearm said, "Hope you hit the jackpot out there, buddy."

The jetty was at the north end of the boardwalk. As you face the ocean while standing on it, Brigantine Inlet is just to the left. To the right sits a long swath of beach capped by the fabled boardwalk and its gaudy casinos. This night a mild northeast wind was puffing and waves were breaking softly on the end of the jetty. The fluorescent glow from the nearby Show Boat Casino negated the need for a headlamp. When I approached, I met another angler named Bill. He told me he had just landed a nice bass around thirty-two inches. It was Bill's birthday. "My wife asked me what I wanted to do for my birthday, and I told her, 'I want to catch a striper.' "

"Well, happy birthday," I said. "Sounds like you accomplished your goal. Now what? Going home or still fishing?" I said.

"Still fishing, man."

When I asked Bill if he knew about the world-record bass caught from the jetty, he had no idea.

"This is a good spot for big bass, but I'm glad I didn't

tangle with that monster tonight," he said and then walked off.

I didn't leave the Vermont Jetty until I hooked a striper, some two hours later. As the tide came in, more of the jetty disappeared. As I watched water shoot through an opening in the rocks like a miniature Old Faithful, a rat scurried for dry ground. My fish—a schoolie—fell for a chicken-scratch Bomber. When I landed it, I felt a sense of relief and walked back to my motel. I stayed on the beach for as long as possible, finally taking to the boardwalk near the Taj Mahal. Behind the glass doors, a rotund lady in a track suit sat at a slot machine waiting for a bite. Over the loudspeakers, a promo for an upcoming George Carlin show was followed by "White Christmas." I was reminded that it was already mid November.

The next day, the northeast wind intensified to twenty knots. I went back to the jetty, curious to see it in conditions that matched those McReynolds had braved. From behind the dune I could hear the *whomp* as each wave met the end of the jetty. Two surfers scurried past with the hoods on their sweatshirts cinched tight. They had wisely left their boards on the roof of their truck. On the beach, I watched as seven-footers rolled almost the entire length of the jetty. The spot where I'd stood the night before was under a couple of feet of water.

My Atlantic City detour had done nothing but stir up my

curiosity about Albert McReynolds, but it was time to move on. Offshore stripers were moving south, and I needed to fall in step.

||| ||| |||

Five months after leaving New Jersey, I received a phone call from McReynolds. He had intercepted my card at Absecon Bay Sportsman Center.

"This is Albert McReynolds, world-record holder of striped bass."

McReynolds went on the defensive before I could utter a word. He told me how well his family had fared in the Atlantic City tournament and the recent plans to include his fish in the IGFA Hall of Fame. In the background, I could hear his wife and sons prompting him. He invited me to fish with his family in Aberdeen, Maryland, and I accepted. We would meet at the Red Roof Inn and fish the Susquehanna Flats of the Chesapeake from shore. The previous spring, his son had landed a sixty-five-pound striper, and the rest of the family had beached countless thirties and forties.

When I arrived, the woman at the counter informed me that the McReynoldses had been looking for me. "I hear they're an interesting bunch," I said.

"I'm not gonna say a word," she told me.

Before I could drop my bags on the floor, I heard voices

outside my door. "Just knock, Dad." Then a heavy rapping. Outside stood Albert and his two sons, Al Jr. and Tom. McReynolds stretched out his hand. "Albert McReynolds, striped bass world-record holder." I had seen only one photo of McReynolds and it was twenty years old. He had not aged gracefully. His body was stooped and he was missing the majority of his front teeth. Life on the road, I imagined, was not easy. Before I could finish saying hello, McReynolds launched into a litany of striper catches his sons had made and then invited himself into my room.

In the twenty years since Albert McReynolds pulled up to Corky Campbell's tackle shop with his fish on the roof, his life has been a series of ups and downs. It's the downs that really get to him. "When you break a world record, your feet leave the ground and you're on cloud nine with your achievement. You don't think people are going to be cruel or nasty or thieves. You're leading with good faith, and you expect them to. It doesn't work that way. When money enters into it, it changes everything."

It was the fighting over money that ruined McReynolds's relationship with many of his family members and friends. "There are many times we were refused for holidays," he said. "People that I thought were my best buddies tell us they have other plans when we visit." McReynolds also said that the $250,000 offer from the hunting lodge dried up when his fish, unbeknownst to him, was thrown away because it had freezer burn.

McReynolds is most galled by the fact that he's considered a pariah along the Striper Coast. "I was at a bar in Maine once when a guy started talking striped bass. I said, 'Hey, man. I'm the world-record holder.' He says, 'You're a goddamn liar, and if you open your mouth again I'll break your jaw.' I bowed out politely and left. That stuff happens all the time."

When I asked McReynolds about those who say it was too rough that night to fish from a jetty, he threw up his hands. "It was rough, and I did get washed off. But the idea that you couldn't fish the jetty is just bullshit. Of course you're not gonna go out where a momma is gonna break on your head and put you through the rocks. We just went out far enough to cast into the wash behind the waves. It's no big deal if the tidal surge breaks on your knees. I grew up on Vermont Avenue and worked as a lifeguard there. I knew that jetty better than anyone."

He took a deep breath. I could tell he'd heard that question before. "You know," he said, "I was just doing something I loved. I can't read or write, but if I could I'd publish a book called *The Night I Hooked the Devil.*" I mentioned that he did win $250,000 for catching a fish, a prize unheard-of even today.

"I've been blessed, man. Yes, I won the richest prize in the history of fishing at that time. And we did live like kings while it lasted. I've met people I never dreamed of meeting— Joe DiMaggio, Ted Williams, Burt Reynolds—all because I caught a fish. But does the good outweigh the bad? I still haven't decided. I'm still shell-shocked from the experience."

I'd gone looking for McReynolds a bit skeptical. I was hoping to find the one crack that revealed the truth, something in his story that didn't jibe. What I found was a simple man who loved to fish, a simple man thrust into a situation beyond his comprehension.

"It's been twenty-years. Do you think anyone will ever break your record?" I asked.

"I know there are one hundred–pound fish out there or better," he said. "It's just a matter of time before somebody catches one."

"What if it's you?"

"I'll take a picture and then cut the line."

11

Delmarva: The Great Brackish Mother

The Delmarva Peninsula (so called because it contains all of Delaware and slices of Maryland and Virginia, in that geographical order) is bookended by the Chesapeake Bay to the west and the Atlantic to the east. Robert Wilson wrote in 1876 that the peninsula takes the shape of the head of a duck, with the neck "constricted between the mouths of the Susquehanna and the Delaware, while the bill seems puddling about among the grass-flats and oyster-beds of the lower Chesapeake." Part of the coastal plain that begins in northeastern New Jersey and stretches to Texas, most of the land is flat and divided by rectangular farm tracts. On the highways, one gets the impression that a chunk of the Midwest has broken loose and been set adrift. Fall comes to these southern latitudes with a gentle nudge; often the appearance of harvesting machinery gives notice to residents before the weather.

I took leave of New Jersey in Cape May by hopping a ferry to Lewes, Delaware. On that late-November day, the ferry, designed for hauling summer crowds, was empty. It would have been possible to play a game of touch football on the auto deck.

With New Jersey behind me, so was the traditional Striper Coast. From here south, on the ocean beaches, the striped bass, which is more commonly called rockfish in these parts, is a bit player, showing only during the spring and fall migrations. During these times, they mix with a plethora of game fish, including the big, bruising red drum. The beaches are flat and wide, the water murky with microscopic sea life and detritus from vast marshes. If you are standing in knee-deep water, you can't see your wader boot. With most anglers favoring hunks of bait over lures, striper fishing, like much surf fishing in the South, takes on an air of reverie. Even the intense blitzing slows until the fish reach Oregon Inlet on the Outer Banks. It's as if the migrating fish take a giant breath to prepare for the final push. Like the fish, the pace of my trip slowed a bit in Delmarva.

I found that many of my contacts had turned their backs to the sea and were watching the sky for ducks or scouting the woods for deer. I zigzagged across the peninsula for a few days, taking in the sights and testing the waters, before arranging a trip with Michael Doebley, the legislative director of the Recreational Fishing Alliance and an avid striper fisherman. Doebley lives in Philadelphia but was on the road drumming up support for his organization. I met him in Rehoboth Beach, Delaware, which fronts the Atlantic a few miles south of the Delaware Bay. We planned to fish the Indian River Inlet and nearby sod banks along the Indian River Bay.

The morning was gray and wet, with a northwest wind urgently tugging flags, as if to tell them it was time to come down for winter. As I stood on a sod bank overlooking a large mud flat, the rain drummed on my tightly cinched hood: *splat, splat-splat-splat, splat.* An outgoing tide had drained the flat of much of its water, and I had little hope of catching a striper, but I'd learned not to trust my initial instincts until I had covered an area thoroughly. We worked our way toward an osprey-nesting platform, casting as we walked. Atop the osprey's plywood platform, litter of every kind—plastic bags, six-pack rings, a doll's leg—was interwoven among the nest of sticks. A dog's leash dangled from the nest, as if somebody's little Petunia had been snatched up and devoured. In the wind, the leash jumped back and forth. On the ground there were numerous large clumps of crab shells that had been picked clean. "Those birds should be eating bunker, not crabs," said Doebley.

To all types of sea life, including whales and game fish, bunker (also called menhaden, and a member of the herring family) are swimming nuggets of protein, the equivalent of Popeye's can of spinach. "The ecological role of the menhaden cannot be overstated," declares *Bigelow and Schroeder's Fishes of the Gulf of Maine*, the seminal text on the fish of the North Atlantic. "They convert energy derived from phyto- and zooplankters, and possibly also vascular plant detritus, into hundreds of thousands of tons of fish flesh." Striped bass relish this flesh. Smaller bass feed on the juvenile bunker (peanuts), while large cows gorge on the adults, some of which weigh more than two pounds. The most storied blitzes in striper-fishing history have occurred when a school of cows intercepted a pod of adult bunker. One such blitz occurred on Martha's Vineyard on Columbus Day of 1981. The chosen few who found this action hauled in forty- and fifty-pounders until they could take no more. Locals still talk about that special day. Many worry they may never see the likes of it again. While the bass are back, the bunker have been conspicuously absent. The result is a population of stripers that look as if they've been placed on a diet. Behind their characteristically huge head is a long, thin body.

Historically, the bunker population, like that of the rabbit, is cyclical, but the cycle hasn't spiked in decades and those connected to the sea have begun fingering another culprit: commercial reduction boats. These 170-foot ships use spotter planes to locate massive schools of bunker, then launch smaller purse boats that circle the school with huge nets. With the fish penned in, the bottom of the net is drawn shut like a purse. Next, the mother ship arrives and sucks the bunker into its hold via a giant rubber hose. When the hold is stuffed with fish, the ship returns to its home base.

Bunker, however, don't end up behind the glass counter in the seafood section of your local supermarket. To the human palate, they are oily, bony, and just plain tough to swallow. Instead, the fish are processed. The fish meal that cattle and poultry are fed is ground bunker. The oil that's left after the bunker are pressed has been used in paints, leather-tanning products, cosmetics, and printing inks which is why some newspapers used to smell like fish when wet. The oil is also rich in omega-3, fatty acids that reduce cholesterol for the millions of heart-healthy Americans who purchase it in liquid or pill form. The most viable commercial fishery on the East Coast is also devastatingly effective. Catch numbers have been declining since 1946, and 2000 was the second-

worst season in sixty years. Critics claim the industry is decimating the fish—once so ubiquitous that early settlers
used frying pans to scoop them out of the water.

An EPA study in Rhode Island's Narragansett Bay focused
on the changes in the diet of the striped bass over the last
two decades after reduction boats depleted the bunker fishery there. What the scientists found, by studying carbon
isotopes in their scales, was that the stripers were forced to
feed on less nutritional invertebrates, like sea worms. To a
human, it's the equivalent of eating a head of lettuce
instead of a steak. Studies in the Chesapeake have also con-
firmed that the loss of bunker has made the striped bass
more susceptible to diseases, such as *Pfiesteria*, a phytoplankton known as "the cell from hell." But the bunker do
more than feed the ocean—they clean it.

A single bunker gulps seawater at a rate of seven gallons a
minute, ridding it of nutritious phytoplankton and detritus.
A school of three thousand fish purifies 1,260,000 gallons
in an hour. As one biologist noted in an article in *Discover*
magazine, "Think of menhaden as the liver of a bay. Just as
your body needs its liver to filter out toxins, ecosystems also
need those natural filters." Without bunker mopping up
this minute plant growth, a window is left open for algae
blooms when excess nitrogen, in the form of fertilizer runoff
or wastewater, is flushed into the sea. The resulting bloom
eventually settles on the bottom, suffocating the life that it
lands on.

Doebley became so infuriated by the reduction boats from Virginia swooping into nearshore Jersey waters and siphoning up the baitfish that he helped form an ad hoc group to fight the raping of the resource. The Salty Dogs of War, along with a host of other environmental groups, fought for legislation to push the bunker fleet out of New Jersey state waters. "The name went over like a lead balloon at our first public meeting, so we became the Salty Dogs," said Doebley. "But we did resort to guerrilla tactics."

The group drafted letters that were spread via the Internet. Each week they targeted a member of the New Jersey Assembly Agriculture and Natural Resources Committee, the agency that was slowing down the progress of the bill to push the boats farther out to sea. "Some guys put their fax machines on auto and blitzed them twenty-four hours a day until we got some recognition." The group also staged a march in Trenton. On a cold, rainy Monday, more than one hundred marchers—three carrying a full-size coffin bearing the message "The Death of New Jersey Fishing"—and a dozen beach buggies advanced on the capital. "It was an issue that galvanized the fishing community," said Doebley.

When we met, Doebley was nearly certain that the barking of the Salty Dogs, alongside the other groups, would pay off. This bill was waiting to be signed by the governor. It was in part because of his early work on this campaign that Doebley was offered the position of legislative director of the RFA.

With the tide draining the flats dry, we headed back to the truck and drove to the Indian River Inlet, which is flanked by two jetties. On an outgoing tide, water hustles from the Indian River Bay through the narrow chute of rocks, taking a host of baitfish with it. Migrating stripers with an urge to eat need only gather at the inlet's mouth during the ebb. On this day, we could see darkened patches of water of what looked like schools of shad getting carried seaward.

Waves thundered up the south jetty, but a break in the clouds let in some sun. Between the jetties, a boat rode the swells. Three anglers jigged bucktails, occasionally rearing back on a striper. The stripers were stacked in the middle of the channel, but Doebley and I could not reach them.

Doebley had a meeting on Long Island that night and left around two in the afternoon. Not long afterward, I took a wave that punched my waist and backed me to the edge of the jetty. It seemed to leach my motivation.

Back at the truck I flipped my calendar open. Thanksgiving was two days away. With three weeks left in my journey, I decided to make a quick break for home—my real home, Savannah, Georgia. Ten hours later I was enjoying a bourbon with my old man.

Rejuvenated after a couple of days of rest and family, I headed to the southern tip of Delmarva to fish the mouth of the Chesapeake Bay.

Cape Charles, Virginia, was the most beautiful run-down town I'd seen on my trip. Its mild state of disrepair seemed suited to the season. Colorful leaves bunched in the nooks of sagging Victorians, the autumn light soft on the over-grown ball field and faded-out storefronts. On the map it's the last town of any size before land's end. But it wasn't big. As I drove around, I counted a total of six cross streets and seven avenues. The avenues, all named for prominent Vir-ginians, led straight to the Chesapeake Bay. I followed Mason Avenue west to see my first Chesapeake sunset. Though I've always preferred sunrises, jaded by the cliché marketers have made of sunsets, this one stopped me. Its swath of light seemed to stream directly into Cape Charles's small harbor, causing the windows of nearby houses to look as if flames roared inside. The silence of the scene was bro-ken by the squawk of terns, and, shading my eyes, I saw a small blitz just offshore. I decided to let this one pass with-out racing for my rod.

In the early 1900s, I would have been called a "come here" by the Cape Charles locals. The tag referred to the never-ending stream of pasty-white northerners who arrived on the train seeking sun and salt air. A planned community, Cape Charles was laid out in 1884 by railroad magnates Alexander Cassatt and William Scott, who were looking for a new terminus for their line that began in New York. Leaving nothing undone, they also established a ferry, which ran across the mouth of the bay to Norfolk, Virginia, and built large, beautiful homes. The town prospered for more than half a century, receiving nearly two million visitors a year. But after World War II, the train stopped carrying passengers. Then, in 1954, the ferry dock was moved south of town, eliminating the need for anyone to stop in Cape Charles. The final blow came with the construction of the Chesapeake Bay Bridge-Tunnel in 1964, which routed traffic around the outskirts of town. These days, fishermen and bird-watchers (who come to get a rare glimpse of the super secretive black rail that lives in the surrounding marshes) provide much of the tourist traffic.

And that's just the way Charlie Stant likes it. Stant, who comes from a family of baymen that goes back to the 1700s, has no need for trinket shops and traffic tie-ups. He lives off a dusty shell road on the seaside of Cape Charles. I met him on a stunningly blue November morning at 9:00. Though the light seemed too thin to hold much warmth, it was T-shirt weather. A week earlier, it hadn't been so pleasant. A three-day cold snap had sent bait running, and many stripers had followed.

I found Stant gassing up his boat. He wore duck boots, jeans, and a button-down shirt. His face, which was anchored by a strong, sharp nose and a thick, white mustache, was etched by a lifetime on the water. "Hop aboard and let's go fishing," he said.

As we motored out of the small tidal creek, we passed Stant's duck blind. This would be his last fishing trip of the year. After a busman's holiday to chase bonefish in the Bahamas, Stant would start guiding duck hunters. During the year, he also crabbed and farmed clams.

His boat, a Chincoteague scow, was built by a neighbor more than twenty years ago. It had a blunt bow and shallow draft, a necessity on the seaside of the lower bay. "It's like a minefield back here," said Stant. "The place is loaded with bars, and none of them are marked. Helps keep people out. It's totally pristine, the way fishing ought to be."

Like many workboats on the bay, Stant's had a tiller stick instead of a wheel. The oak stick came up to just above his waist. It was worn as smooth as a church pew. When he pushed forward, the boat turned left; pulling back sent it to the right. "The stick's great for snaking through guts and when you're working crab pots. You can drop the stick and pull your pot over and there's no console to get in the way."

I was taking a few notes when I felt Stant rap me on the head. When I looked up, he pointed his chin to the horizon. A massive skein of black ducks lifted from the water. For a moment the flock blotted out the sun.

Stant eased up on the throttle near the point of a marsh-grass island where numerous gulls rested on the water. A few loose feathers on the surface meant they had probably been wheeling about in a frenzy over feeding fish minutes before. "Looks promising to me," said Stant, with understated conviction. Through the fiberglass hull, which acted as a megaphone, we could hear a colony of snapping shrimp as they repeatedly closed their tiny claws at breakneck speed, producing a sound not unlike popping corn. In the open ocean, large colonies often make such a racket they jam the sonar of navy subs. "They look a lot like miniature lobsters," said Stant. We both made a cast with our fly rods to an eddy created by the sweeping current. My leader snapped when I set the hook, but Stant's striper stayed connected.

The two of us each pulled in half a dozen stripers before the school moved on. "Time to give civilization a shot," said Stant, referring to the Chesapeake Bay Bridge-Tunnel.

The Bridge-Tunnel, as locals know it, loomed a mile or two distant. It stretches 17.6 miles across the bay's mouth. Its 5,114 concrete pilings attract fish of all types. In the late fall, if the water in the bay is warm, a fair chunk of migrating striped bass make a right-hand turn at the mouth of the Chesapeake and congregate there for the winter.

Since we carried only fly rods, we planned to fish some of the shallow water nearby. I had dreamed of dipping a finger in the mouth of the Chesapeake for as long as I'd been a striper fisherman. While the Hudson River produces striped bass, 80 percent of the migratory stock comes from the Chesapeake. The majority of the fish I'd followed had, at least once, swum beneath the bridge on their way to the ocean. If Jack Kerouac called the Mississippi River the "great brown father of waters," then the bay was the great brackish mother of striped bass.

Ellington White, in his story *Striped Bass and Southern Solitude*, used a different but equally evocative metaphor for the Chesapeake. "Stripers seem to regard the bay as a school they have to complete before graduating into the Atlantic Ocean. The school lasts four years. A few dropouts may tackle the ocean sooner than that, but the majority are content to wait until graduation day. Then they are ready to join the big ocean community on the outside."

Mother or school, the mouth of the bay yielded no fish. The sound of trucks thundering along the bridge was enough to make Stant wince, and we headed back to the seaside. He slowed down at the southernmost tip of Delmarva, where a flock of turkey vultures perched on wooden pilings as if they'd been raised by pelicans. At least twenty of them sat fidgeting, like schoolchildren on a class trip, and squawking. They were waiting on a favorable wind to make the crossing over the bay and continue south. As one of them sent a squirt of shit flying, Stant shook his head and hit the throttle.

Back at dock, Stant pulled out two ice-cold Cokes and we sat in wooden chairs beneath the shade of loblolly pines that shot skyward like arrows. Stant seemed in no rush to close his last striper trip of the year, and I was happy to oblige. We swapped fishing stories, but I was eventually trumped by Stant's tale of a one hundred–pound tarpon that literally jumped in his scow one afternoon while he was hauling crab traps. Before he could subdue the wildly thrashing fish it made a shambles of his boat. He had a picture to prove it.

In front of us, the setting sun gilded the water in the small creek. At our backs, fields of pine alternated with farmland. As the languid afternoon in Cape Charles inched toward dusk, I realized I was in no hurry to become a "leave here."

North Carolina: The End
of the Striper Road

By the time I reached the Outer Banks of North Carolina, roadside flags had become my wetted finger to the breeze. Figuring wind direction was easy, but I also became adept at gauging speed. I'd known since childhood that a full-size American flag snapped straight in a twenty-knot breeze, but it seems only car dealerships have the patriotic fervor for giant Stars and Stripes. Smaller gas-station and supermarket flags—the type that push product, not freedom—

require some freelance thinking. You have to take into account the way they yank at their halyards or bend their flimsy poles. After three months on the road, I could guess the wind within three knots. Sitting at a stoplight on my way into Nags Head, I watched a bright-green flag that advertised a BP gas station rip from its mooring and blow right across the highway. This was a new one on me, but I put it at thirty knots. I turned my weather radio on—twenty-eight knots.

The wind, which had started blowing on Cape Cod and had pestered me since, kept me motel-bound for two days. Each morning, I would wake up and walk to the beach, where sheets of sand would blast me, stinging any exposed part of my body. On the way back to my room, I had to spit sand grains out of my mouth. With my money dwindling, I located a few campsites where I could pitch my tent if need be. I had reached the end of the road—I would surely wait out the wind.

Thirty years ago, my trip would have ended at the mouth of the Chesapeake Bay. In those days, many anglers knew that a portion of the migratory stock wintered in the waters off of the Outer Banks, but these fish escaped with little fishing pressure. In his 1966 book, *Secrets of the Striped Bass*, Milt Rosko wrote, "An angler with time on his hands and a yen to try something different could easily stumble on some virgin fishing in this broad expanse of wilderness."

News spread in the early 1970s when a *Sports Illustrated* cover showed a haul-seine crew on the Outer Banks surrounded by a pile of huge striped bass, but not long after the stock collapsed and the waters around the Outer Banks grew silent in the winter.

III   III   III

Devin Cage was there the day the stripers returned to the Outer Banks. It was a November morning in 1993, and Cage was heading offshore to do some commercial fishing for king mackerel. As is normally the case with the North Atlantic in the fall, conditions were horrible, and the boat turned back before leaving Oregon Inlet. But through the clear water, Cage and his buddies noticed the black backs of fish in the inlet and threw some cobia jigs to them. "Nobody carried striper lures anymore, but the jigs worked," said Cage. The group pulled in a boatload of fish.

Since then, big striped bass have been arriving at Oregon Inlet every November. It's the only time that concentrations of Hudson River fish mingle with Chesapeake Bay stock, but it's easy to understand why they come. The shallow, nutrient-rich waters provide a home for dozens of fish species, many of which end up in the stripers' gullets. Along with the herring and bunker, which have also migrated from the north, croakers, weakfish, trout, flounder, and other edibles spend their lives in the waters off the Outer Banks. "The rockfish arrive kind of thin," said Cage, using the colloquial name of the striped bass, "but they get a lot bigger down here. Then by January we get a push of really big fish—fish in the forties and fifties." By March, the fish begin migrating back up north to the spawning grounds.

When the wind eventually let up, I arranged to meet Cage at his charter boat at 6:00 A.M. He was booked for the day but had no problem with me tagging along as an observer.

When I showed up, it was still dark, but a yellow glow came from the cockpits of the sportfishing boats lined up against the dock at the Oregon Inlet Fishing Center. Mates readied the day's tackle, a ritual they performed by rote. Larger boats were headed offshore to chase tuna, while the smaller boats would stay near the beach and target stripers. With his three clients aboard, Cage motored his forty-two-foot boat through Oregon Inlet and backed it down to within casting distance of the sandy shoals that make this slip of a waterway one of the most dangerous on the East Coast. Since 1961, twenty-three people have died while running the deadly broth of current, waves, and sand.

The inlet is relatively new. Its genesis occurred during a hurricane that battered the Outer Banks 115 years ago. Tremendous waves breached the dunes and set a torrent of water loose from Albemarle and Roanoke Sounds. After the storm passed, a deepwater inlet remained, the only one along the 220 miles of coast between the Chesapeake Bay and Cape Lookout, North Carolina. The first ship of any significance to run the inlet was the steamship *Oregon*. The waterway was named in its honor. As is nature's tendency, the inlet was far from static. Its channel seemed to whip back and forth, like a hose gushing water. But not only were the shoals constantly migrating—a vessel spending more than twenty-four hours at sea could expect a drastically different channel on return—but the inlet itself was moving. At first the pace was hardly discernible, but in recent years the waterway has moved south to west at an impressive rate of 313 feet a year. It now stands two miles from the Bodie Island Lighthouse, which was erected in 1872 in part, to mark the inlet's entrance.

At the confluence of the Labrador Current, barreling down from the north, and the Gulf Stream, chugging up from the south, the inlet is essentially a washing machine, stirring sediment, depositing it, stirring it again. Solid shoals of sand move around like snowdrifts. Add to this the swells from the North Atlantic, which roll in from offshore and, when their bellies feel the friction of sand, rise up and blot out the horizon, and the inlet can be a death trap. The week before I arrived, it was just that. Allen Frucci, a sixty-eight-year-old retired lieutenant colonel in the U.S. Marine

Corps, died when his boat most likely capsized on the shoals surrounding the inlet.

An avid striper fisherman, he was reported overdue by his wife when he failed to return from a morning fishing trip. Frucci, who had flown 320 combat missions in Vietnam, was the most decorated Marine Corps pilot in the war; his honors included the prestigious Flying Cross. He was a man who had stared fear in the face. And that, one local charterboat captain told me, was what may have killed him. "He would run those shoals in all kinds of weather. He wasn't afraid of it. We would see him out there in his seventeen-footer when none of us would have touched it. He was a hell of a skipper, but too much confidence can kill you out there."

Conditions on the day Frucci died were far from ideal. Fog limited visibility, and large eight- to ten-foot swells spawned by a distant hurricane bombarded the area. At 4:48 P.M., Frucci's wife alerted the Coast Guard that he had been expected home at noon but had not returned. The search began immediately. Beach crews spotted debris in the water, including lifejackets, and a rescue craft was launched. Frucci was found floating facedown about four hundred yards from shore. The next day his capsized boat was recovered.

Cage was well aware of these dangers when he dropped his boat back into the cauldron. His three clients sat in chairs bolted to the deck as his mate cast live eels into the white water. The pitching deck was a challenge as the boat bucked in the mishmash of waves. It was prime striper habitat. Within thirty seconds, two rods were bent, and Cage motored out into more gentle seas. Once the fish hit the deck—both twenty-pounders—Cage circled and put the boat right back in the fury.

"What a fishery—this is just amazing. This is world-class," said John Whitehurst Jr., the leader of the group. Whitehurst, who buys and sells timber, lives in nearby Currituck on a spread of land that teems with deer and other wildlife. If Whitehurst and his gang had any fear, they didn't show it, even when a rogue wave rolled over the port gunnel, drenching the mate and the youngest guy in the trio, who yelled, "This is just like *The Perfect Storm!*"

With five nice stripers in the box, Cage decided to ease out of the maelstrom and troll the outer edges of the shoals. "I trust Devin," Whitehurst told me. "When the fish are hot, he gives me a call. What he says goes." As the mate dropped the jigs back, Whitehurst opened a cooler and pulled out a grilled venison tenderloin, tomatoes, bread, and a host of condiments. "Help yourself," he told me. "I'm not happy with the marinade I used on the venison, but it's not bad. And the tomatoes are the last of this year's crop." Even though it was just past 8:00 A.M., I made a sandwich. Whitehurst's knife, which I used to slice the tomatoes, was disconcertingly sharp, especially in a rocking boat. Whitehurst may have been unhappy with the marinade, but it was some of the best venison I'd ever tasted.

Whitehurst and his buddies had driven in that morning and would leave the Outer Banks by the afternoon, but the relatively new striper fishery already draws anglers from all over the East Coast. "These fish," said Cage, "have created a winter tourism industry for the Outer Banks."

It was easy to see why. By 9:30 the boat had reached its limit of eight stripers, the last three having been caught while trolling with up to three bucktail jigs on each line. When all four rods went down at once, Whitehurst called for me to take one. I pulled in two fifteen-pound stripers. "What an unbelievable fishery," Whitehurst hooted. Suddenly, the water all around us was roiling with silver flanks of giant stripers. There was such commotion it was difficult to tell what unlucky species of fish was being devoured, but when the stripers sounded, millions of tiny scales floated on the surface like a sprinkling of glitter.

With more than a dozen other fish released, Cage decided to pull the lines. He had another party waiting to leave the dock at noon, but if he got them on the boat early he would have a good chance of making it home before his kids got out of school.

"You're more than welcome to come on the next trip with me," said Cage, "but I bet those fish will be on the beach any minute now."

"I was thinking the same thing," I told him as I packed my gear. On the dock, Whitehurst and his gang posed with their catch. "Stop by my land this afternoon. We'll put you up in a deer stand and you can take some venison home to New York. We have plenty of deer," said Whitehurst.

"I would love to, but I have a few more fish to catch before this adventure ends," I said. "And from the looks of it, this might be as good as it gets."

III   III   III

The Outer Banks are known for redfish. The world record, a ninety-four-pounder, was pulled from the waters off Avon in 1984. Redfishermen, much like their striper brethren to the north, travel the beaches in trucks loaded with rods and frequently spend the night on the sand. The two fish are similar in many ways, except redfish don't blitz, and they aren't too fond of artificial lures. Most angling for them is done with chunks of cut bait. When the stripers returned to the Banks, few fishermen pursued them with lures. Things have changed since then, but you'd be hard-pressed to find an Outer Banks fisherman with a lure arsenal like a Yankee angler's.

As I crossed over the Oregon Inlet bridge, I noticed a swarm of birds hovering over the breaking waves. From that distance they looked like a ball of gnats. On the shore, a few fishermen were casting to them. I gunned the truck and sped off to a turnout near an abandoned Coast Guard station at the south end of the bridge.

I grabbed my surf bag and rod and ran through the soft sand. After two hundred yards, I slowed enough to catch my breath, then kept chugging. I could see a few bent rods and humps on the sand that were stripers. Every fisherman on the beach was using a big Kastmaster with a white rubber screwtail on the hook. I followed suit and waded into the water.

Just outside the breakers, gannets dive-bombed the ocean. One after another they dropped from the sky like the needle of a sewing machine, patching a hole in the sea left by giant feeding stripers. Mixed in were pelicans, which looked like lumbering transport planes compared with the stealthy gannets. While the gannets landed with little commotion, like Olympic divers, the pelicans came down with a thud, then surfaced with fish flopping in their fleshy jowls. It reminded me of a saying I used to hear from a charterboat captain when I was a kid. "Mr. Pelican, Mr. Pelican. Your mouth can swallow more than your belly can."

My first cast was unproductive. So were my second, third, fourth, and so on. Occasionally someone next to me would let out a modified rebel yell as he set the hook on a fish. I hadn't seen one under fifteen pounds hit the sand; most were in the twenty-pound class. This was the type of blitz fishermen remember during winter nights—it was a season maker—and I couldn't buy a bite. I left the tight group of fishermen and ran south toward some more bird activity. I told myself to relax. I wasn't doing anything wrong. My lure just hadn't been in front of any fish. In front of me the water erupted, and I made a quick cast. Nothing. Others joined me, and soon more damn rebel yells. If I'd had a musket, I may have let loose with a volley.

For thirty minutes I went fishless. Frustrated, I ripped open my plug bag. I grabbed a Hab's needlefish: no reason it shouldn't work down south. When the guy next to me hauled his fish in, I sent the lure seaward.

When I tightened up on my line, the needlefish nosed its way to the top of a wave, and just behind it lurked a striper, like a dog sniffing a soup bone. Excited, I jerked the lure away from the fish, but before I could curse myself, another striper grabbed it. Ten minutes later I beached her—twenty pounds. From then on I could do no wrong. Even when there were no fish to be seen, stripers lunged from the water to take my plug. I heard the guy next to me yell to his buddy, "Damn, he's catching them on a top-water lure. I told you those might work here." Moments later they were running to their truck. They returned forty-five minutes later, ripping the packaging off brand-new surface plugs. I was on my knee releasing my sixth striper. At forty-two inches— probably more than thirty pounds—it was the largest of my trip. She was bright, with a greenish back and dark black stripes. In the water, I rocked her back and forth, which pushed water through her gills. I felt her regain her strength, watched her dorsal fin rise up, and then with a powerful flick of her tail, which sprayed water in my face, she swam away. Afterward I hooked my lure to the eye of my rod, content just to watch. As I sat down on the sand, I let out my own rebel yell.

||| ||| |||

That night I blew my remaining cash on a room at the Holiday Inn Nags Head. It had a balcony overlooking the Atlantic, which I stepped onto around 11:00 with a beer in hand. Standing there, I thought about the first school of big bass I'd encountered at Higgins Beach, Maine, three months earlier. Somewhere in the dark ocean below me, chances were good that they were joining other similar-size fish in a massive school. For the rest of the winter, they would patrol the sandy shoals and cool, deep waters of the Outer Banks, always on the lookout for their next meal.

Sometime in late February, however, an urge stronger than hunger would stir inside of them. They would swim up the tributaries of the Chesapeake and spawn. Afterward, spent and tired, the fish would head north to the joy of anglers gone loco with cabin fever. Their arrival would be met with a collective sigh of relief. Another season would begin.

The bass, of course, have no idea of the delight they carry on their shoulders for nearly four million striper fishermen on the East Coast. We fishermen simply impose ourselves on their existence. That night, looking out over the ocean, I decided I had imposed on the stripers long enough this season.

The next day I drove home.

Acknowledgments

Writing a book is a leap of faith. Jump without a safety net and you'll land with a thud. My net included a handful of friends who made this a better book with their expert advice and suggestions. A few who went beyond the call of duty are Jenny Comita, Sid Evans, Will Palmer, and David Willey.

Before hitting the road, Jim Butler, Wilson Laney, Joe Malat, John Mazurkiewicz, and Tom Rosenbauer provided me with much needed contacts. Kathleen Hamilton at Ford helped me land a suitable vehicle. On the road, I met more people than I could squeeze into two books, much less one. A few people in the book will notice our considerable time together was condensed into a scene or two of salient events. I did this to keep the pace consistent and the reader engaged. I've also kept my promise, in a variety of ways, to keep secret spots hidden. Some individuals who aren't in the book deserve special mention: Bill Cole, Roland Fujimoto, William Hoxter, Bill Hubbard, Lev Huntington, Tom Keer, Bob Luce, David Ross, Rich Tenreiro, and the Long family.

Much credit for this book goes to the two men who made it a reality. My agent, David McCormick, thought it was a good idea from the start and, most importantly, sold it. My editor, David Conti, bought it. His deft touch is evident throughout the book. Thanks to Leslie Falk and Knox Huston, who both put up with the niggling questions and concerns of a first-time book writer.

I'm lucky to have the pages of this book graced by the work of Keith Witmer, who captured the essence of the Striper Coast with his wonderful line drawings.

I'm thankful that my first mentor was a man named Bill Gerken. He taught a young kid about the wonders of a tidal river—and the best ways to catch fish from it. I couldn't have been luckier.

Even closer to home, literally, I'd be remiss not to mention my uncle, Joe DiBenedetto, who introduced me to the striped bass on that foggy night in Long Island Sound. He's been a trusty fishing partner since. I'm grateful to my brother Bob and his wife Catherine, who entrusted me with Robert for a week. My brother Christian, a computer wizard, talked me through technical difficulties at all hours of the day and night. And last but certainly not least, I'd like to thank my parents. In so many ways, none of this would be possible without them.